A GUIDEBOOK *to the* GEOLOGY *of* *Lake Superior's* APOSTLE ISLANDS NATIONAL LAKESHORE

and Nearby Areas of the Bayfield Peninsula of Wisconsin

Revised Edition

BY EDWARD B. NUHFER

Poetry by Mary P. Dalles

EASTERN NATIONAL • FORT WASHINGTON, PENNSYLVANIA

This guidebook is dedicated to Dr. Dwayne Stone, "Stoney," Professor Emeritus, Marietta College, who brought the joys of geology and good instruction to so many students; to Robert and Leone Dalles; to Chad and Sandie Dalles; and to the memories of Edward Philip, Mary Agnes Nuhfer, and Dr. William Norton McNerney.

Front cover image: Caves in the Devils Island Sandstone at Swallow Point on Sand Island
Back cover image: Stack from the Chequamegon Sandstone

Originally published by the authors, 1987
Revised edition published by Eastern National, 2004
Reprinted by Eastern National, 2006
Fort Washington, Pennsylvania 19034

ISBN 1-888213-95-7

Library of Congress Cataloging-in-Publication Data

Nuhfer, Edward B.
A guidebook to the geology of Lake Superior's Apostle Islands National Lakeshore and nearby areas of the Bayfield Peninsula of Wisconsin / by Edward B. Nuhfer; poetry by Mary P. Dalles.– Rev. ed.
p. cm.
Includes bibliographical references and index.
ISBN 1-888213-95-7
1. Geology–Wisconsin–Apostle Islands National Lakeshore–Guidebooks. 2. Geology–Wisconsin–Bayfield Region–Guidebooks. 3. Apostle Islands National Lakeshore (Wis.)–Guidebooks. 4. Bayfield Region (Wis.)–Guidebooks. I. Dalles, Mary P. II. Title.
Q180.A66N85 2004
557.75'21–dc22

2004008032

Contents

Lake Superior

Lake Superior is a woman–
do you know–
holding her children
Cat, Rocky, Madeline
and the others
in a strict love
but only as mothers do
to bring out beauty
and strength.
Are not the sea caves
evidence?

Lake Superior is a woman–
do you know–
soothing her children
Hermit, Outer, Michigan
and the others
in a gentle love
with long fingers
caressing the sandstone
and the red cliffs
in times when the sun rises
from the center of the earth
and falls back into it.

Lake Superior is a hundred women,
all the myths of woman,
the ancient hag who has endured
a billion years;
the virgin child new and untouched
clear, pristine;
enticing lover, gently decorated
for your pleasure
with ruby sunsets and pearl of moon;
eternal fathomless essence of woman
and who knows whether this be
or is merely dreamed.

Lake Superior is a woman
you can too easily
fall in love with,
one of those women with whom
living is a risk
but a glorious one;
too much intimacy may destroy you,
but oh to meet the five-foot wave
in a bass boat
and match strength with strength–
what sweet exhilaration.

But Lake Superior is not a dream–
do you know–
though she may contain your dream;
and she waits as the dream
and as the woman
to slip into your knowing,
to brighten the nighttime
of your days.

Preface

Geology is a science of materials (rocks, minerals, water, air), processes (weathering, erosion, transportation, deposition, tectonic processes, etc.) and change through time. Geology guidebooks are often exclusively technical works, written for geologists and geology students to provide sound information on the features found in an area. Certainly, the geology is one of the most striking features of the Apostle Islands. The arches, balancing rocks, wave-cut cliffs and benches, tall bluffs of red glacial till, the beaches and sandspits—all these are geological features. Many people who are not geologists visit the islands; they come for the beauty of the islands and the adventure of exploring them. A visit to the islands permits one to experience the scientific observations of the geologist in a setting fitting to the romantic reflection of the poet. We therefore decided that this guidebook, like the islands it describes, would contain elements of both science and poetry.

These islands contain a wealth of geological features and may represent some of the finest examples of braided stream deposits in North America. They can be as serene as calm summer sunsets or as violent as the storms of their home in Lake Superior. To say "the Apostle Islands are beautiful" does not even begin to capture the experience when a visit here is done correctly. A visit to see the geology of these islands is a combination of learning to recognize different kinds of sandstones, cross-beds (inclined beds of sand deposited by waves and currents), ripple marks, fingerprints of ice-age glaciers, erosional coastal features such as sea caves and wave cut benches and depositional coastal features like spits, beaches and lag deposits. However, it is also a visit with fishermen, sailors, rangers, and photographers. It includes being rocked by waves and sprayed with cold water; getting sand in your shoes; watching a matchless sunset; camping under a full moon; being among gulls, eagles, and cormorants; looking over maps in campfire light; and eventually returning to Bayfield for a dinner of poached trout and hot coffee. This is time spent in the finest way, and it yields rewards of education and adventure that last a lifetime. We, like other visitors, fell in love with the islands; each day spent here brought increased appreciation, and every passing year leaves memories of these visits undiminished.

Each of the twenty-two islands has its own unique character and yields its own special memories. We have resisted displaying features with other than black and white photos out of respect; the visit should provide its own feel and true colors of that day and hour. Such an experience should always be a cut above any book. Whether you have come for the day, the weekend, or the summer, this book should simply enrich your time spent here. The ultimate compliment to our guide is to see a smudged copy with some sand and water spots on it and many wrinkles from multiple trips. We would then know that it was a companion in happy adventures.

Geology augments the appreciation of what one sees in the moment by adding a perceptual vision through which you see the features as the products of processes that act over tremendous spans of time. The feature observed then takes on a dynamic quality that goes beyond its immediate appearance. Understanding geological features seen in the field is akin to learning to read. At first, you see only undecipherable black marks on a page. As such, these surely are not very exciting. However, as you proceed to learn to read, the simple markings reveal endless stories and fascinating events. The same is true in geology. At first, you see mere outcroppings of rock, but as you learn to "read" these, an intriguing story begins to unfold. Satisfaction gained through understanding earth processes is not at all the proprietary territory of geologists. Enjoyment and appreciation are more dependent on interest and attitude than upon mere technical training. This satisfaction can be accessible to anyone.

Along with adventure experienced in exploring these islands comes intimate contact with the Earth and its processes. Here, a single day may be spent in walking upon rock formations that have persevered over a billion years of time, in boating along cliffs left by glaciers ten thousand years ago, and in feeling the cold, pure waters of Lake Superior wash over the sands on which we wade today. Part of the wonder lies in knowing that today is the product of hundreds of billions of days. The sands and sandstones have witnessed them all and seem to say, "We are still here!" Here and now, we can reflect upon the place of our own lives within this vastness of land, water, and time. "What is truly important?" reverberates through the mind, not so much as a question but as a thunderclap.

Introduction

Access to Features

The Apostle Islands (Fig. 1) are accessible from commercial tour boats, sailboats, and powerboats of varying sizes. In the recent ten years, these islands are a particular destination of sea kayaks. For current tour boat routes and shuttle services, you should contact the National Park Service headquarters in Bayfield. If you arrive with a personal boat, your first stop should be the headquarters in Bayfield. There, you may get advice on launch sites, weather, location of ranger stations, and obtain useful books and charts. Learn the nature of the area in which you will be traveling before you depart. *A Visitor's Guide to the Apostle Islands National Lakeshore* by Dave Strzok (2003) is available at the headquarters and at several of the local marinas. This is an indispensable reference and should be a companion guide to this book if you pilot yourself around the area. Judziewicz and Koch (1993) also have published a compilation of Apostle Islands' flora.

You may be here for the first time and decide that your first access may be on the tour boat. The tour routes change from year to year, but in all cases, figure 1 will help you locate your position.

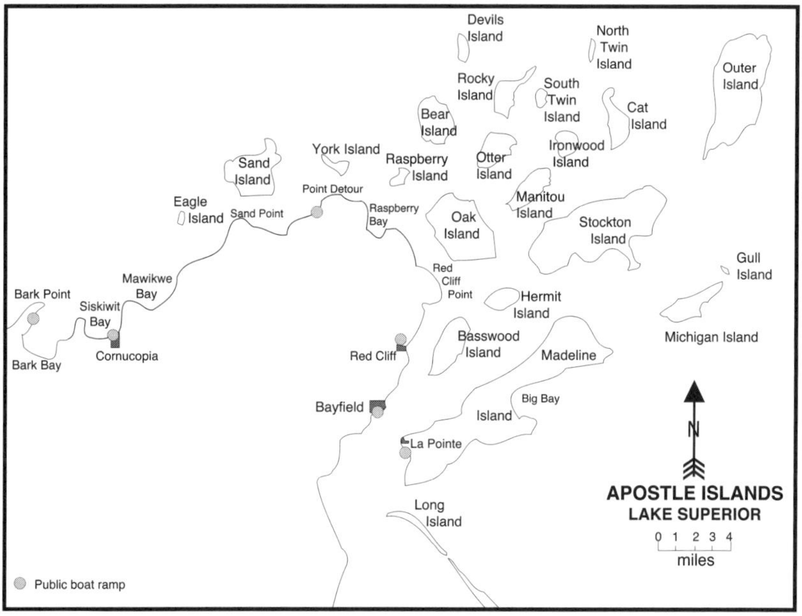

Figure 1. Outline map of Apostle Islands and Bayfield Peninsula showing locations of islands and major sites referred to in this guidebook.

Most of the text designed for use in the field is arranged alphabetically island-by-island. There is much to see on the tour boats, and you will be too busy to read this from cover to cover. Enjoy reading the rest later. Doing so will bring back good memories. If you have time before the boat departs, you may wish to read the section "General Geology of the Islands." On the tour boat, you can make good use of binoculars (7X or greater is recommended), a camera (if you have a zoom or telephoto lens, bring it), and a light rain jacket or windbreaker if you wish to enjoy the weather outside the covered passenger deck. Even though we've boated to each of the twenty-two islands, we still enjoy an occasional trip on the larger tour boat. It was our own first introduction to seeing the islands, and it's a good one. It's safe, inexpensive, and informative. If you choose to utilize your own boat at a later time, a prior trip by the tour boat can help familiarize you with the islands, show you some of the conditions of wind and water you are likely to experience, and reveal many areas which you might like to examine more closely and for a longer period of time.

Access to a number of geological features on the islands themselves requires landing on or skirting close to the shore. This requires a small boat. Pilots of large vessels that cannot venture near shore or land in rocky areas will find it useful to carry a small boat or inflatable raft in tow. In this guidebook, you will sometimes read passages that refer to piloting a "small boat" into certain areas. Be aware that there are "small boats" such as sea kayaks, that are very seaworthy in these waters, and that there are other "small boats," especially flat-bottomed johnboats, tiny car-top boats, and canoes, that are unstable in waves and not safe for travel between islands or even along some mainland shores. Distances between islands are sufficiently great that a slow boat or paddle-driven canoe caught between islands or too far from shore cannot escape to safety when Lake Superior waves come up quickly (and typically!) in a gusty wind. When you arrive at the National Park Service Headquarters in Bayfield, a ranger will be able to look at your boat and recommend safe limits for your boating activities. Be certain to listen and act accordingly with that advice. None of the stories you have heard are remotely bad enough to approach the experience you will have if caught in weather beyond the capabilities of your boat. The nearest shoreline is not necessarily shelter. Some shorelines of the park and the nearby areas are sheer rock cliff that cannot be climbed and lack any place to land a boat. In the cold offshore waters of Lake Superior, situations that might require one to swim a long distance or even remain afloat in a life jacket for hours are life-and-death situations.

When planning your trip to the islands with a small boat, give yourself plenty of time for your visit. It may take you only twenty minutes to get to Sand Island, but if the weather changes, and it often does so in a matter of minutes, your trip can extend to several days until calm weather allows your return. Plan that any open small boat will travel light, below its rated capacity of weight and number of passengers. Ratings for small boats assume use on calmer rivers and small lakes and not on an inland sea like Lake Superior. Your boat, when loaded, should be fast enough to allow you to arrive at your destination under the same good weather conditions under which you depart. About eight miles-per-hour is the minimum that will allow inter-island travel within time spans of less than thirty minutes. Good life preservers and extra gasoline should be included in your equipment.

As a final reminder, your first stop should be the National Park Service Headquarters in Bayfield, to obtain safety information, advice, and weather updates. For small boats that aren't of the size routinely used for inter-island travel, we have outlined one-day trips in more protected areas that show excellent geological features (see "Sheltered Water Tours").

F. T. Thwaites made the first detailed geological studies of the Apostle Islands in a small boat named the Oronto (Fig. 2). The Wisconsin Geological and Natural History Survey sponsored his expeditions (Thwaites, 1912). The survey director (M. Ostrom, 1986, personal comm.) noted that Thwaites often referred very affectionately to his boat, an affection quickly understood by readers who have piloted their own crafts to many of the islands. The Oronto was not a typical small boat. It was an exceptional craft designed by an engineer for use on Lake Superior by only one passenger plus gear. We did our fieldwork for this guide in a fourteen-foot modern aluminum "bass boat" with an extra-wide beam and an extra-deep hull, similar in size to that of the Oronto. Unlike Thwaites, we found we could not travel at will, but often had to remain camped on an island for days until suitable weather presented itself.

Figure 2. The Oronto on the Lake Superior beach in 1912. The wooden boat weighed only 250 pounds, could carry three times its weight in cargo, and could travel at seven miles-per-hour with an efficiency of twenty miles-per-gallon. Such performance is good even by modern standards. (Photo courtesy of Wisconsin Geological and Natural History Survey.)

Sailboats and sea kayaks are the most popular recreational vessels in this area. If you are here with a sailboat, you have both the safety of a respectably large boat and the accessibility afforded by a dinghy or rubber life raft. This allows you to enjoy specific features and study them in detail. If you are here for the first time with a sea kayak, your experience will leave no doubt about why they are so popular here. One can proceed silently very close to shore and perceive every detail that sight, sound, and touch can offer. One truly feels at one with the lake and the islands when present in a sea kayak.

The World in a Bass Boat

A submerged boulder can rip out
the bottom of your bass boat;
ramming it,
you can be thrown into waters
sometimes 505 feet deep,
lethally cold.
Tooling along at 18 horses
is flying;
Lake Superior passes under you
in gigantic flowing blocks
determined by how much water
you can scan faultlessly
in a measure of the eye.
Sustained focus wipes out time,
becomes the agent through which
vital powers are exercised.
Light is
only as it is filtered through water,
as it discovers,
as it assists your discovering.
As the sun falls,
the lake becomes a solid thing,
darkly impenetrable,
and you search it now
not because of any perceived need
although there is need,
but because not to look ceaselessly
at the waters–
smooth and black perhaps as a polished eggplant,
lurching perhaps like a rodeo bronco,
sometimes rippling in a herringbone tweed–
is to give credence
to a less pure reality,
to leave the pristine recesses
of inside of mind.

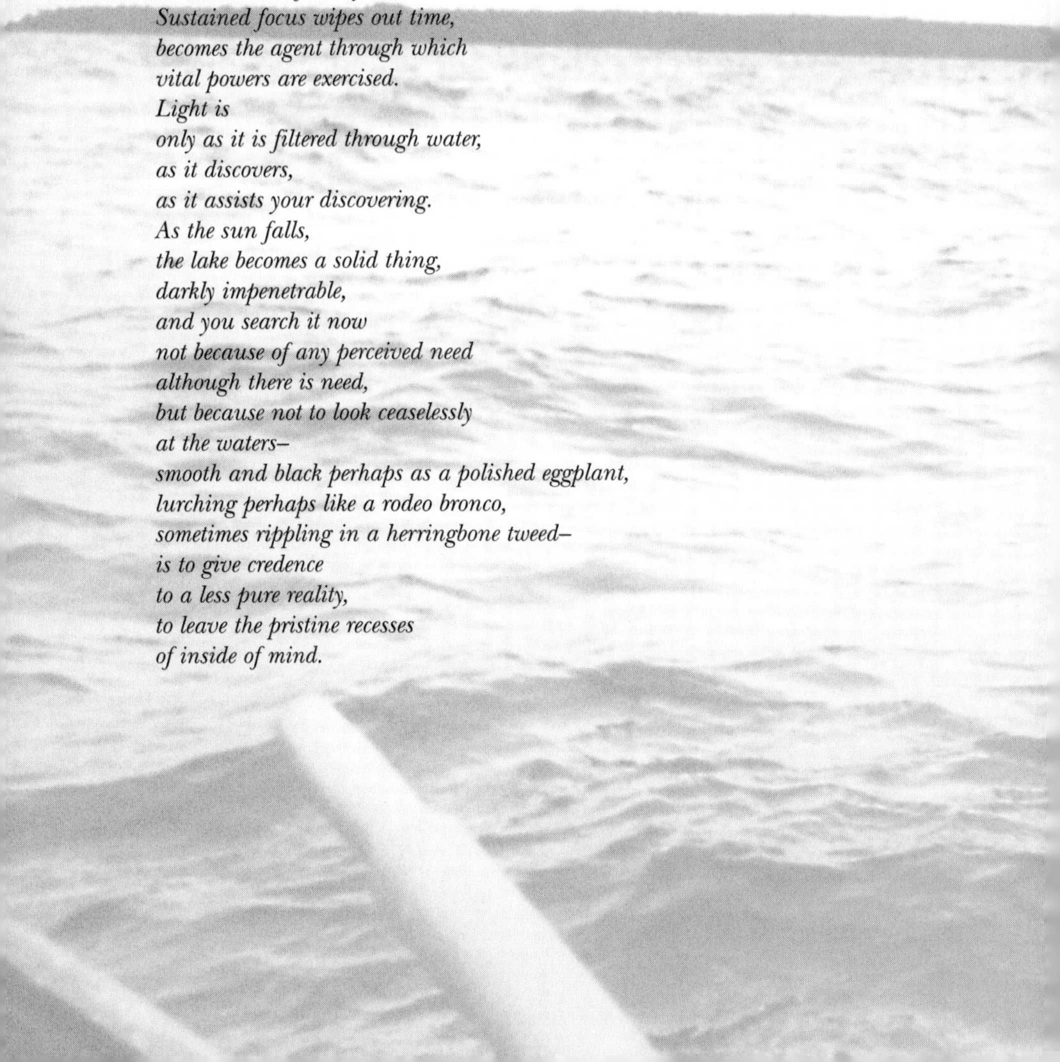

General Geology of the Islands

There are three main aspects of the geology of the Apostle Islands. The first is the geology of the one-billion-year-old bedrock—the geology of the solid sandstones from which the beautiful sea caves, balanced rocks, stacks, wave-cut cliffs, and platforms are formed. The second is the glacial geology—the geology of the 10,000-year-old deposits left by the last advance of the Pleistocene continental ice sheets. These form the high reddish bluffs and the clay banks, which provide soils for vegetation and serve as the main sources of sands for the beaches and sandspits enjoyed by many visitors. Finally, there is the geology of the present, the processes of interaction between water and solid earth that sculpt the bedrock and construct the beaches and cuspate forelands (see Fig. 57). The degree to which all three aspects are offered within the islands makes the area a virtual treasure for the outdoorsman, the naturalist, and certainly the student of geology.

Bedrock Formations and Early Geological History

The bedrocks of the Apostle Islands are sandstones that are divisible into three formations. To a geologist, the term formation has a very special meaning. It means that the rock contained within the formation is actually a unit that can be mapped and traced over some distance. Formations derive their names from published rules termed "The North American Stratigraphic Code" (North American Commission on Stratigraphic Nomenclature, 1983). All formations take their names from specific geographic features such as a town, an island, a mountain, or some other feature actually labeled on topographic maps. The original description of any formation must include a description of a type locality, which is a detailed description of an extensive exposure near the namesake locality that shows all of the mappable features of the formation. These mappable features include the rock type, its color, its texture, its thickness, and distinguishing features such as the manner in which it weathers or an abundance of some particular sedimentary structure (for example, ripple marks). Finally, the author must publish his/her description along with his/her proposed formation name in a valid, peer-reviewed journal. Geological maps that display formations are literally road maps to the subsurface (Fig. 3).

Several associated formations are sometimes lumped together in a formal unit called a group. The acceptance of group names is also contingent upon meeting specified requirements in the Stratigraphic Code. Likewise, individual formations are occasionally subdivided into members. The United States Geological Survey (USGS) compiled accepted names of groups, formations, members, and their type localities and physical descriptions in print form in a volume called *The Lexicon of Geologic Names of the United States* (Keroher and others, 1966). New formation names are

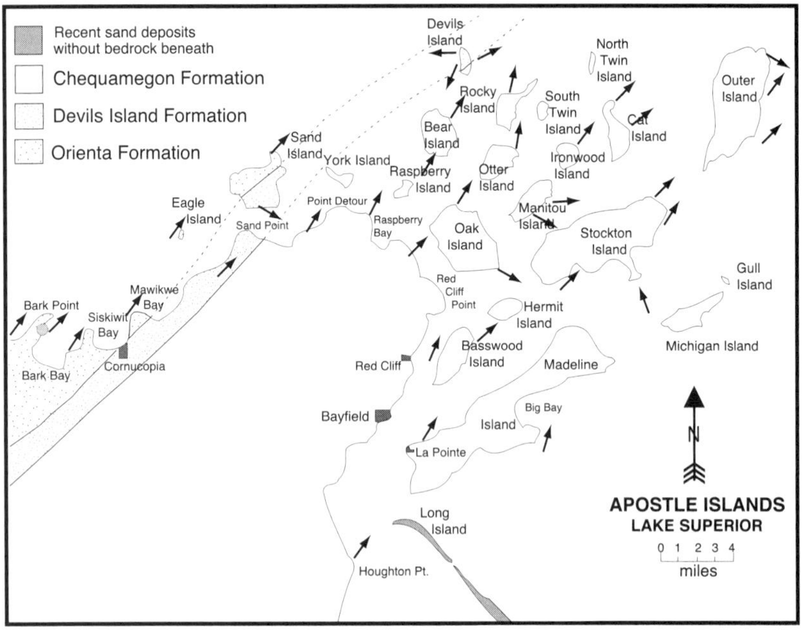

Figure 3. Geological map of the Apostle Islands showing the outcrops of the three Precambrian sandstone formations (after Thwaites, 1912). The sandstones are not perfectly horizontal, but instead are inclined ("dip") a few degrees to the southeast. This dip allows exposure of all three formations at the surface within the national lakeshore.

still being assigned and revisions published. Fortunately, these and the Stratigraphic Code are now accessible in a searchable online version as the USGS National Geologic Lexicon at http://ngmdb.usgs.gov/Geolex/.

The rocks that underlie these islands consist of three formations in a group known as the Bayfield Group. Thwaites (1912) defined the Bayfield Group as consisting of, in ascending order (from oldest to youngest rocks), three sandstone formations: the Orienta Formation (Orienta Sandstone), the Devils Island Formation (Devils Island Sandstone), and the Chequamegon Formation (Chequamegon Sandstone; see Fig. 3). The Chequamegon Sandstone was named for its type locality at the stone quarries on Houghton Point just north of Washburn on Chequamegon Bay; the Devils Island Formation, for its exposures on Devils Island (Fig. 3); and the Orienta Sandstone, for its type locality in a quarry near the town of Orienta on the west side of the Bayfield Peninsula. The USGS National Lexicon still recognizes the Bayfield Group and its three formations. Later studies of this bedrock and the basin in which it formed were done by Leith, Lund, and Leith (1935), Tyler and others (1940), Abel (1985), Ojakangas and Morey (1982a, 1982b), McCabe and Van der Voo (1983), Davidson (1982), Halls and Pesonen (1982), Hinze, Wold, and O'Hara

(1982), Kalliokoski (1982), Morey and Green (1982), Luetgert (1982), Van Schmus, Green, and Halls (1982), Morey and Ojakangas (1982), Wold, Hutchinson and Johnson (1982), Watts (1979), Myers (1971), White (1966), Mudrey, and Ostrom, (1986). Morey and Van Schmus, (1988) and Reed, (1993).

The Area Before the Deposition of the Bayfield Group

Many workers have wrestled with unraveling the details of what happened in this area over a billion years ago. The problem has been a difficult one, and perhaps a complete story may never be compiled. However, we do know that one of the major events pertinent to the islands was the formation of the Lake Superior Basin itself. This began about 1.2 billion years ago with an event that created the Midcontinent Rift. Geologists believe that the massive plate of rock that now constitutes the core part of North America actually began to split or separate. This was not a catastrophic event, but took place slowly over tens of millions of years in a manner similar to the present separation of rock in eastern Africa along the East African Rift. As the plate began to split, the crust thinned in the area of the break, and the area was active with volcanoes. Fracturing of the crust led to large outpourings of lava and emplacement of related intrusions. Although the rift extends for more than 2,500 kilometers northward from Kansas beneath Lake Superior and Michigan, surface exposures of rocks related to the rift occur only in the Lake Superior region. The force that acted to form the rift eventually subsided and the rift ceased to separate farther after about 1.1 billion years ago. The crust began to sag in places in small basins, which eventually were to grow into the larger Lake Superior Basin. The volcanic rocks were uplifted and eroded, and their eroded debris were deposited mainly as the sands found today in sandstones of the Oronto Group. As rifting waned, a huge block of older granite, which may have stretched from Duluth southwest through Minnesota and into Iowa, remained in high relief at the western edge of the rift valley. After streams eroded the overlying volcanic rock, they cut on downwards into the granite. Streams carried sand from the highlands northward and eastward. Younger sands buried the older sands of the Oronto Group, and these younger sands constitute the Bayfield Group seen here in the islands. Thus these sandstones, which are currently believed to be about a billion years old, nevertheless constitute the youngest rocks in the area. Table 1 summarizes the early geological history of this area.

SEQUENCE OF PRECAMBRIAN EVENTS AFFECTING THE APOSTLE ISLANDS

About 1 BY ago	*The northeasterly-flowing streams deposit Bayfield Group of sandstones. Northern Wisconsin lay near the equator at this time. Lake Superior Basin develops into large single basin.*
About 1.1-1.0 BY ago	*Small basins continue to coalesce to form a larger Lake Superior Basin, but granitic rocks in an upthrown fault-bounded block (termed a "horst") remained uplifted in the area of what is presently Minneapolis-St. Paul, Minnesota (the St. Croix Horst). These rocks are eroded, and northeasterly-flowing streams carry the sediments from this source area to the Apostle Islands area. Midcontinent Rift becomes inactive.*
About 1.2-1.1 BY ago	*The basalts are eroded and sediments yielded by their erosion are deposited as sandstones. Rifting peaks at 1.1 BY ago. These sandstones, the Oronto Group, underlie the Bayfield Group.*
About 1.2 BY ago	*Volcanism and extrusion of basaltic lavas ("Keweenawan volcanics"—basalt is a black igneous rock rich in iron and magnesium) occur in developing rift. Lavas accumulate in small basins.*
About 1.4(?) to 1.2 BY ago	*The Midcontinent Rift opens and the Lake Superior Basin begins to develop as a series of small basins.*
About 1.6-1.8 BY ago	*Period of uplift occurs as well as some volcanism. Metamorphosed granites are exposed and eroded.*
Prior to 2.5 BY ago	*Granite crust forms and granites later undergo regional metamorphism.*

Table 1. General sequence of events that affected the Apostle Islands during the Precambrian, listed in ascending order from oldest to youngest. BY = billion years. This table summarizes the early rifting, volcanism, uplifts and erosion that led to eventual deposition of the Bayfield Group sandstones.

The Origin and Nature of the Bayfield Group of Sandstones

The Bayfield sandstones were once sands carried by rivers and along lake beaches. The actual age of the Bayfield Group has been debated as between about 1 billion years (late Precambrian) to 600 million years (early Cambrian), but the most recent evidence still favors the older age. In these times, there would have been no terrestrial vegetation; hence, the rivers would have flowed through valleys, sandbars and channels that were totally barren. The streams that deposited the sands apparently had their headwaters in a highland to the west and southwest, the eastern edge of which stretched roughly from Duluth to Mankato, Minnesota. Granite appears to have been the main rock exposed in those highlands at this time. Granite is a rock comprised of the minerals orthoclase feldspar (potassium aluminum silicate) and quartz (pure silica). Sandstones that are rich (>25 percent) in feldspar grains are termed arkose. The Orienta and Chequamegon sandstones contain only about 11 percent and 14 percent feldspar respectively (Myers, 1971), which would make them subarkose in the formal classification of sandstones. Feldspar breaks down more easily than quartz, so sands worked over by streams for a long time gradually have the feldspar worn out of them, and the sands become increasingly quartz-rich. Sandstones made of more than 95 percent quartz are called quartzose sandstones. Devils Island Sandstone contains about 99 percent quartz, and the underlying Orienta and overlying Chequamegon formations contain about 83 percent to 85 percent quartz (Myers, 1971). As the granite eroded, tributary streams carried the quartz and feldspar-rich sediment from the highlands to the lower parts of the rift valley. These streams apparently entered the rivers, which in turn carried the quartz-feldspar sands toward the northeast (see Ojakangas and Morey, 1982a; Morey and Ojakangas, 1982). Portions of these river deposits are exposed in the Apostle Islands.

With respect to the Bayfield Group itself, the northeastward-flowing streams deposited the oldest of the three formations (the Orienta Sandstone). The Orienta Formation is about 1,800 feet (575 m) thick and is exposed along the western portion of the Bayfield Peninsula, around Bark Bay and Siskiwit Bay near Cornucopia, and on Eagle and Sand Islands (Fig. 3).

Eventually, sand carried by the streams began to fill the basin into which they flowed, and a lake surrounded with sand beaches formed in the area of the Apostle Islands. The area probably consisted of many shallow lakes and ponds, some only a few inches deep, all lying upon a vast expanse of fine sands. These sand deposits constitute the middle formation of the group, the Devils Island Formation. It is about 300 feet (90 m) thick, and exposures may be seen on Devils Island, on Sand Island, and along the western portion of the Bayfield Peninsula along Mawikwe Bay (Fig. 3).

After some time, the basins subsided again, and streams resumed their flow across the Apostle Islands area. These streams, like those that deposited the older Orienta sandstones, flowed from southwest to northeast. Eventually, stream sediments again blanketed the region. These sands were buried, cemented and became the Chequamegon Sandstone, which is about 470 feet thick (150 m) and the uppermost formation of the Bayfield Group. The Chequamegon Formation occurs on the majority of the islands of the national lakeshore (Fig. 3) and is exposed along the Bayfield Peninsula and on Madeline Island.

There are many types of stream deposits, but based on the strongly dominant northeast paleocurrent direction, the abundance of sand in the rock, the scarcity of clay, the thickness of the sand beds and occasional beds of gravel (conglomerate beds) found in the rock, we (Nuhfer and Dalles, 1987) believe the streams that deposited the Orienta and Chequamegon formations were a type of stream called a braided river. Braided rivers are usually shallow; their channels are only a few feet deep. They carry a heavy load of sand, and soon their channels choke with sand, diverting the channel elsewhere. The Platte River in Nebraska and the Kicking Horse River (Cant, 1982) in British Columbia are probably good modern analogs to these deposits. Within Wisconsin, the portions of the Wisconsin River near Boscobel and Muscoda that contain many sandbars would have some similarity (see also Figure 4).

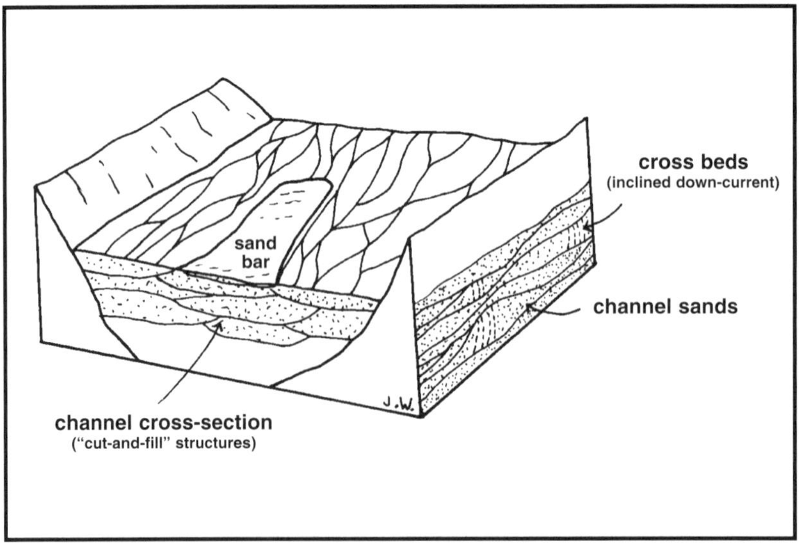

Figure 4. Block diagram schematic showing shapes of sandstones deposited by braided streams. Note that the individual sand beds are mostly thin and that the beds undulate in thickness, are more continuous when viewed in direction parallel to stream flow, and show most channel cut-and-fill structures when viewed at right angles to direction of stream flow.

The characteristic features of braided streams are displayed as well in these Precambrian rocks as in many modern braided stream environments. Sedimentary features that we will illustrate with photos later in this guide (planar cross-beds, thin channel sands with scour cross-beds, convoluted bedding, parting lineation, scour-and-fill structure, and thin gravel deposits along with mudcracks and ripple marks) are very well exposed and accessible to the boater. These features are what geologists use to decipher both the paleoenvironment of sandstone deposits, and the directions in which the streams carried the sands.

Figure 5 shows a paleocurrent map of the streams that deposited these billion-year-old rocks, as deduced by study of these sedimentary structures. Note that the islands have a general southwest to northeast trend or "grain." A few of the island shapes that vary significantly from this trend also reveal bedrock whose paleocurrents depart from that trend. This is probably where the streams meandered a bit, just as do present-day braided streams.

The exposure of the Bayfield Group through erosional processes is precisely what makes the Apostle Islands area of such interest to geologists

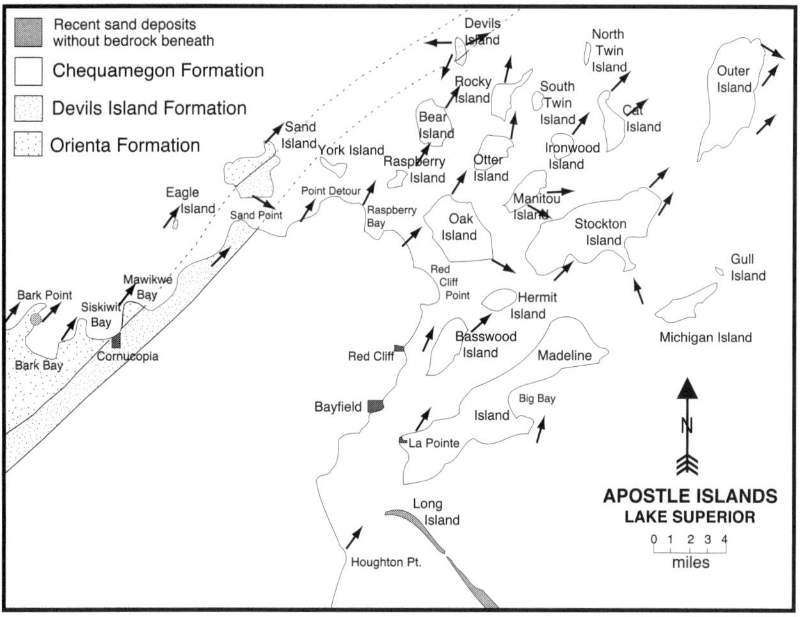

Figure 5. Paleocurrent flow directions (arrows) as deduced from measurements (mostly from Myers, 1971) on directional sedimentary structures. The directional trends are more diverse and scattered in the Devils Island Formation owing to an increase in wave-formed structures rather than current structures, which dominate the Orienta and Chequamegon formations. Outcrops viewed from east or west will be mainly views of beds parallel to stream flow direction; outcrops at northern ends of islands expose beds cut perpendicular to stream flow (study also Fig. 4).

and what provides the rugged cliff scenery appreciated by all of us. The erosion of the overlying layers in combination with the sculpturing by lake water and by glacial ice has yielded beautiful three-dimensional exposures of features on several islands. In some places, the bedrock has been stripped in such a way that you can actually walk upon the surfaces of old sand flats, river sandbars, and river channel bases that preserve an event that happened on a few days over one billion years ago.

This region was later covered by seas that deposited marine sediments, but these sediments were eroded away through millions of years of exposure. None of these marine sediments remain here. To see these sediments, you must proceed eastward to the area of the Door Peninsula and on into Michigan where the downwarped Michigan Basin allowed several thousand feet of these to accumulate and be preserved.

The "commercial mineral deposits"—iron, copper, nickel, gold, and silver reported in 1973 (Matteson and others, 1973)—remain undocumented in the literature. It is likely the authors mistakenly associated mineral deposits found within older, much different igneous rocks to the northwest and southeast with these sandstones. The only documented commercial geologic deposits associated with the rocks of Apostle Islands National Lakeshore were industrial minerals, the "brownstones," that were quarried here in the late 1800's. Many exquisite buildings made of brownstone, one of which is the Apostle Islands National Lakeshore Headquarters in Bayfield, remain standing today. Further discussions on the brownstone industry occur in later sections of this book.

After the Bayfield sandstones were buried, they were folded into upwarps called anticlines and broad downwarps called synclines. The rocks are not quite flat-lying in the islands, but instead are slightly tilted to the southeast at an angle of about 2° to 5° on the westward margins of a broad syncline. When rocks are folded even slightly, stresses create a regional pattern of fractures called joints. Joints are nearly vertical in their orientation and usually appear straight when viewed in a cliff. Although jointing produces definite breaks, there is little or no shifting of the rocks on either side of the break (Figs. 9 & 64). Within a region, joints usually occur in a group called a "joint set." From a map view, joints within the same set have essentially the same parallel compass direction or trend. It is the usual case for a region to have two dominant sets. The two sets that are dominant in the Apostle Islands trend about North 15° East (essentially parallel with the elongate trend of the outermost islands) and North 80° East. As you shall see when visiting the islands, the regional joints are important in the formation of sea caves and stacks. If there were significant movement along a break, the break would be termed a fault instead of a joint. There are no major faults known in the park.

The Glaciers and the Apostle Islands

The Pleistocene Epoch, informally known as the "Ice Age" because of the continental glaciers that then covered presently temperate areas, encompasses an interval of time from about three million years ago up to about ten thousand years ago. However, the Pleistocene is just one of many such ice ages. Continental glaciers here were similar to the glacial ice cap that now covers most of Greenland; they were extensive, thousands of feet thick, and fluctuated periodically in size because of changing conditions in climate. Several possible causes of occasional expansion of the polar ice caps to form continental glaciers have been proposed. One of the more accepted ideas is that subtle variations in the Earth's orbit occur in regular long-term cycles (called Milankovitch Cycles after the Yugoslavian geophysicist who proposed their existence in the 1930's). These variations are able to affect climates to a degree that may permit significant expansion of continental ice sheets. Whatever the explanation, there is no doubt that glaciers have covered a significant portion of the north-central United States many times. Abundant evidence for glaciation in the immediate region of the Apostle Islands exists for the period between 20,000 and 10,000 years ago. Based on data from a larger region of North America, we know that glaciers invaded this local area repeatedly before then. The most recent glaciers have buried or swept away much evidence of the earlier ice ages here.

When Pleistocene glaciers entered the Apostle Islands area, northeastward-flowing streams had already eroded valleys down into the Precambrian sandstones. Fish Creek near the southwest end of Chequamegon Bay near Ashland, the Sioux River near Washburn, and the Sand River on the Red Cliff Reservation follow older Pleistocene stream courses. The Bayfield Group sandstones probably determined the locations of the Pleistocene streams. Where older Precambrian rivers had established their main channels, they deposited thicker and more massive sandstones, and outside these major stream channels, they deposited finer and more easily eroded materials. As the younger streams continued to erode, they dissected the northerly portion of the Bayfield Peninsula (which probably at that time extended all the way to Devils and Outer Islands). Eventually, the lower (northerly) portions of these stream valleys broadened; interstream divides began to wear down, and the streams flowed within valleys between isolated hills underlain by the thicker, more resistant sandstones deposited long ago as bars and channel fills by the Precambrian streams. Those hills are probably today's Apostle Islands.

Let's turn back the clock about 20,000 years to the time of what is termed the "St. Croix Advance" and begin to imagine the advance of the glacial ice into this area of hills. By examining maps of later advances (Figs. 6 & 7), we can check our imaginations against reality. The ice, hundreds perhaps thousands of feet thick, filled the Lake Superior Basin and advanced from the northeast into this area.

The forefront of the ice entered the valleys first and then continued to build in thickness as the main body of ice followed. Glaciers can remove sediment (alluvium) and even large boulders from stream valleys as easily as a bulldozer moves soft soil. As the thicker ice encroached upon the lower islands such as Michigan, Cat, North Twin, South Twin, and Rocky,

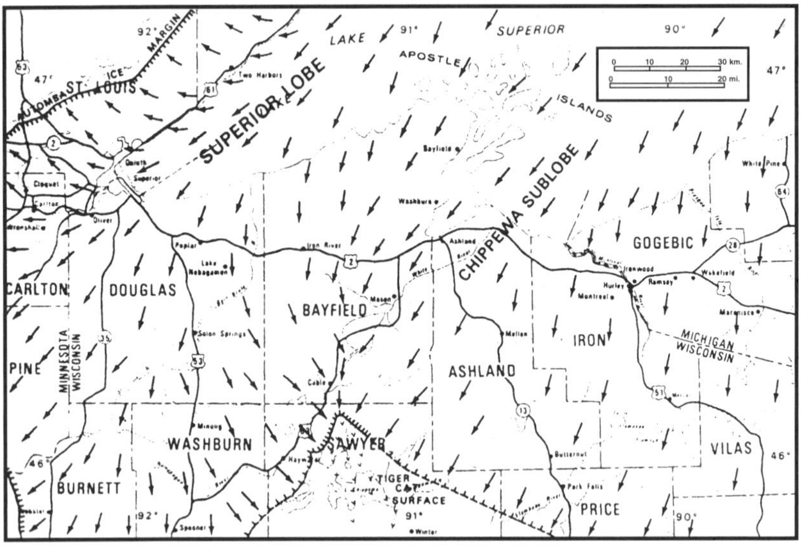

A. *Tiger Cat Advance*

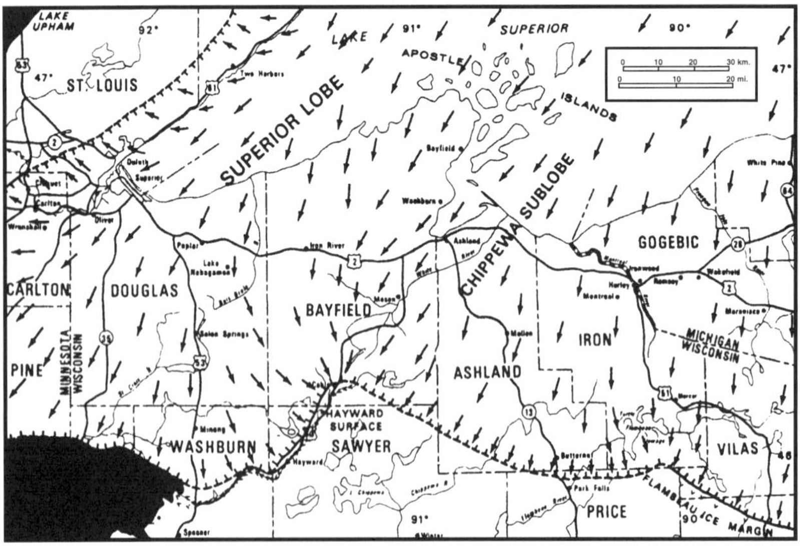

B. *Hayward Advance*

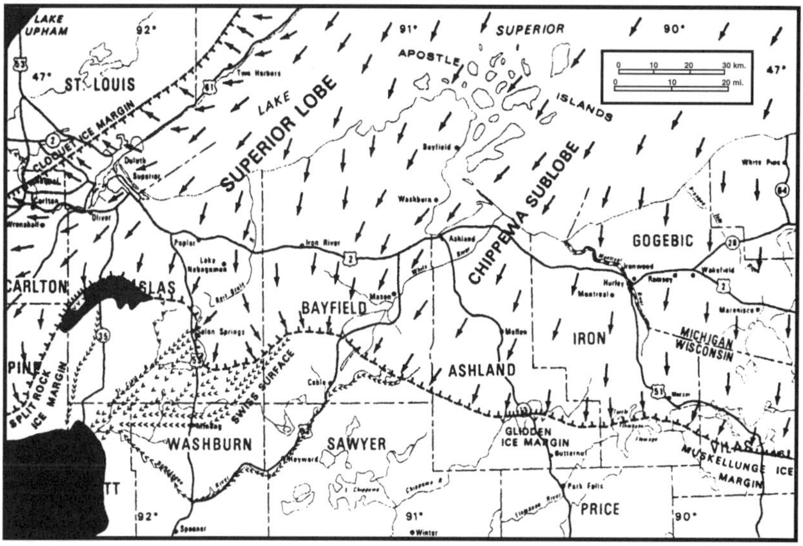

C. Swiss Advance

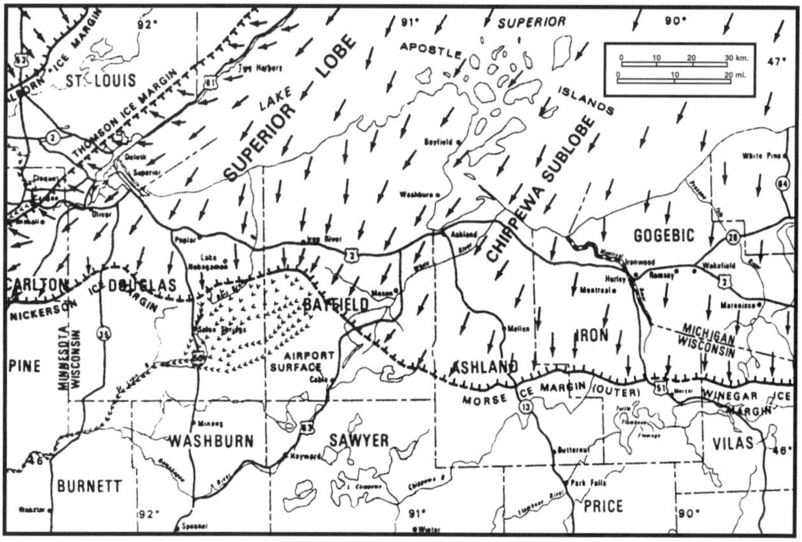

D. Airport Advance

Figure 6. Maps showing general extent of early advances of ice sheets over southern Lake Superior (from Clayton, 1985). Sequence A through D is from older to younger. These advances of ice overrode the area of the Apostle Islands and the Bayfield Peninsula. Arrows show direction of ice advance; small arrowheads show direction of meltwater flow; heavy lines show margin of ice advance marked today by terminal moraine; black areas show sites of ancient ice-marginal lakes.

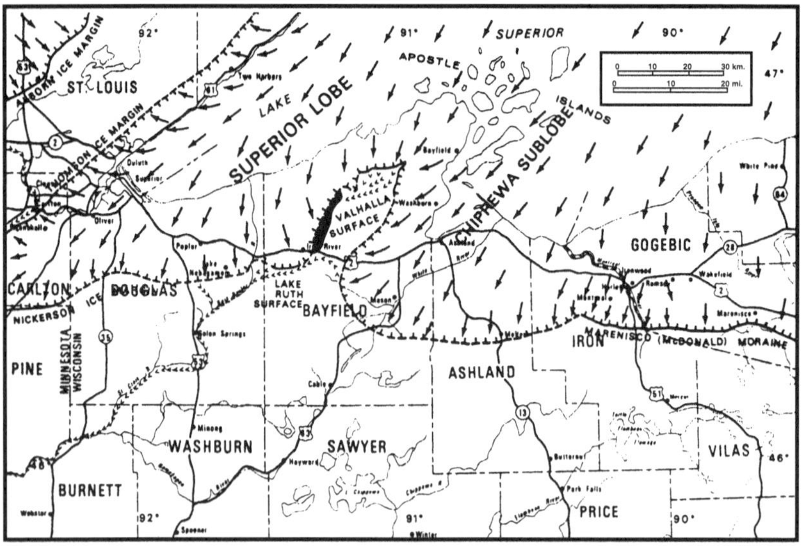

A. Lake Ruth Advance

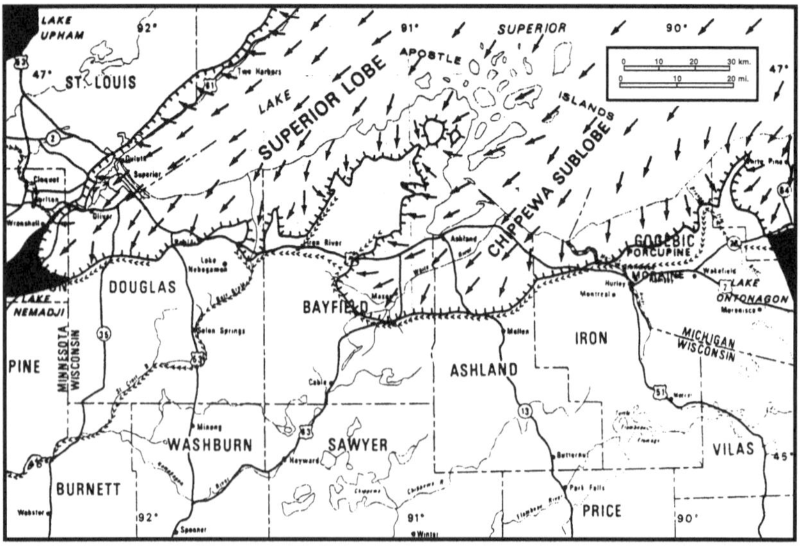

B. Porcupine Advance

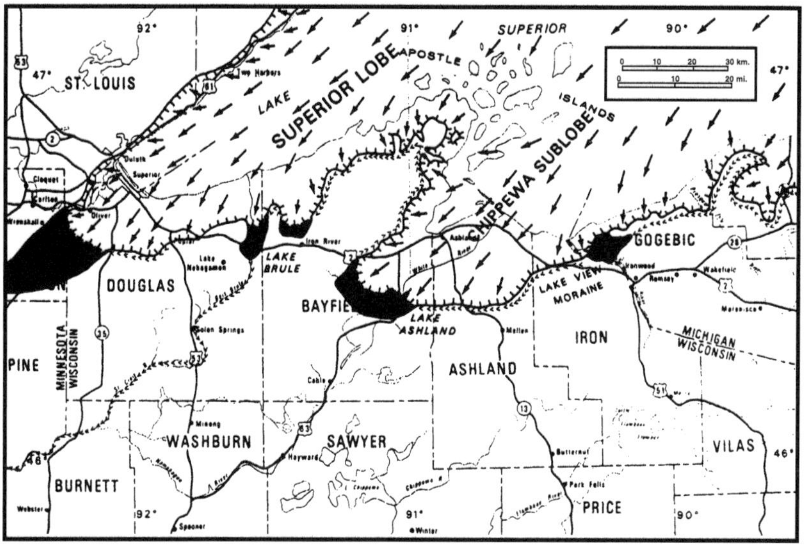

C. Lake View Advance

Figure 7. Maps showing general extent of more recent advances of ice sheets over southern Lake Superior (from Clayton, 1985). Sequence A through C is from older to younger. The Lake View Advance represents the most recent glaciation of this area and occurred a bit less than 10,000 years ago. Note the smaller southerly extent of the more recent ice advances in comparison to those of figure 6. Ice margins now are along the Bayfield Peninsula. As the Lake View Advance later receded, ice margins retreated northward to the area of the islands. Arrows show direction of ice advance; small arrowheads show direction of meltwater flow; heavy lines show margin of ice advance marked today by terminal moraine; black areas show sites of ancient ice-marginal lakes.

it rode over them, stripping away the soil and vegetation all the way down to the Precambrian sandstone bedrock. The boulders that the ice picked up along its journey accumulated at the base of the glacier and were dragged along under the immense pressure of the overlying ice. These cut glacial grooves and striae into the bedrock—grooves that we can observe to this day on many of the islands, particularly on Stockton and Bear Islands. The glacier continued on its way and was split by the highlands of the peninsula and the Apostle Islands (particularly Oak) into two lobes—one which passed into the area that is now Chequamegon Bay, the other which passed to the west of the peninsula along Lake Superior. However, as the main body of thick ice arrived, the southwestward-advancing ice covered all the islands, and finally the Bayfield Peninsula itself.

Climatic conditions changed, the ice began to melt, and the glaciers receded from south to north. But what a volume of meltwater is released from a melting ice sheet that in places is thousands of feet thick! The southernmost edge of the ice served as a dam or barricade to prevent the streams from resuming their old northerly courses, and water ponded up there until it was sufficiently high to begin to flow out over the lower land to the south and eventually cut new southerly-directed stream valleys. This, in fact, happened right on the Bayfield Peninsula. The ice lobes on either side of the peninsula towered over the lower land where the ice was thinner because of the original topography. Meltwaters poured in from the north, west, and east. These meltwaters formed powerful, rapidly flowing streams that carried cobbles and boulders. Such streams often carve distinctive features called potholes in underlying bedrock. Potholes form in rapids below water falls, where the stream churns a hard cobble around in a depression until it carves a deep hole with exquisitely smooth, almost polished, sides. One can see potholes on Sand Island at the western side of Lighthouse Bay just a foot or two below the presently normal lake level elevation of 602 feet (Fig. 56).

As the streams poured in water, a lake formed at the base of the glaciers' southern margin. Eventually, as the waters deepened, they formed the headwaters of a mighty river. The channel began at about Brule, Wisconsin, and flowed southward along the present (much smaller) St. Croix River Valley to the Mississippi River. The channel, termed the "St. Croix Spillway," drained Lake Superior, which then stood at a maximum elevation of about 1,075 feet (328 m—Clayton, 1985), over 470 feet higher than now. This process was repeated (Fig. 6) a few thousand years later in the "Tiger Cat Advance," then the "Hayward Advance," followed by the "Swiss Advance" and then the "Airport Advance," which occurred about 12,300 years ago.

Later advances—(Fig. 7) the "Lake Ruth Advance" at 11,500 years, the "Porcupine Advance" at 11,000 years and finally the "Lake View Advance" at less than 10,000 years ago—were successively smaller and did not completely cover the Bayfield Peninsula (Clayton, 1985).

Each of these ice advances brought with it debris deposited from previous advances, and carried many rocks from the Lake Superior Basin and from Canada into the Apostle Islands area. Therefore, a variety of igneous, sedimentary, and metamorphic rocks as well as some less common minerals are found in the boulders, cobbles, and gravels on the islands themselves and especially along their shorelines (see section on South Twin Island). With the exception of Gull and Eagle islands, and Long Island, which is unique among these islands, glacial till covers most of the exposed surfaces of the islands. Till is simply a mixture of materials of various sizes dropped directly by melting ice. The high reddish bluffs that are particularly characteristic of the western sides of most of the islands (especially Michigan, York, Bear, Oak, and Outer Islands) are mostly glacial till. The lake bottom itself has a thick layer of glacial deposits, as revealed by seismic profiles (Berkson, 1972; Scholz, 1984).

Clayton (1985), Connell (1984), Johnson (1980), Need (1980), and Leverett (1929) have described the tills of the area. Contributions by these workers are the basis for table 2. The most detailed work to date describes the tills on the western side of the Bayfield Peninsula from Wisconsin Point to Bark Point.

TILLS OF THE APOSTLE ISLANDS AREA

Till Name	***Age***	***Description and Formation Name***
Douglas Till Member	*10,200 years old*	*Dull reddish brown till that is clay-rich west of Port Wing and sandier to the east. A member of the Miller Creek Formation. The Miller Creek Formation, possibly the Douglas Till Member, covers the older glacial Copper Falls Formation on many of the islands and appears in many exposures.*
Hanson Creek Till Member	*12,000 years old*	*Reddish brown clay rich till characterized by many gray clay stringers between 1/2 in. and 4 in. thick, present only west of Iron River. Color of this till is slightly less red than the Douglas Till. A member of the Miller Creek Formation.*
Jardine Creek Till Member	*14,000 years old*	*Dark reddish brown sandy till resting in places upon brecciated (intensely fractured) bedrock, likely fractured by the overriding ice. A member of the Copper Falls Formation. Shoreline tills to date that have been mapped lie within the Copper Falls Formation; some are likely Jardine Creek Till Member.*

Table 2. Glacial tills found in the Apostle Islands area. The most prominent till on these islands is the Douglas Till Member.

In some areas, the till was apparently overridden by ice to form a subtle streamlined topography characterized by low elongate hills of till called drumlins, which are oriented parallel to the direction of the latest ice advance. While the casual visitor doesn't easily notice these, they are evident from aerial photographs. Clayton (1985) mapped drumlins on Outer, Cat, Stockton, Devils, Michigan, and Madeline Islands. Occasionally, small ponds or lakes may form within till, and fine clay and silt, typical of sediments in contemporary small lakes and ponds, will fill these. Clay banks such as those seen along eastern Oak Island appear to be old pond deposits.

In addition to till, some glacial material found on the islands is well-sorted sand and gravel—either deposited by running meltwater from glaciers or left behind when waves winnowed out fine clays and silts during times of high water stands. Each ice advance was followed by a high-water stage of Lake Superior as the ice receded and melted. Lake terraces, the elevation of which corresponds closely to the paleoelevations of the lake's waters, mark these high-water stands. Previous workers have defined terraces near the Bayfield Peninsula area at elevations of about 475 feet, 380 feet, 285 feet, 220 feet, 120 feet, and 30 feet above the present normal elevation (602 feet) of Lake Superior. These terraces represent times when the lake remained near a constant elevation long enough to erode a niche at the shoreline. When a lake remains at constant elevation for a time, it produces depositional features like beach deposits, or erosional features, like wave-cut benches. Water-sorted deposits above present lake levels can be seen along the shores of several islands, particularly near the Outer Island sandspit on the east side of the island, and at places on South Twin and Rocky Islands. Erosional benches are less obvious, but a hiker may notice them on Oak Island along the interior trails. There were times when Oak Island or Oak and Bear Islands were probably the only emergent islands above Lake Superior. The well-sorted fine sand that forms the soils at the high point on Oak Island (elevation 1,081 feet) is likely beach sand deposited at the high water "Duluth Stage," which formed terraces elsewhere at an elevation of 1,076 feet.

The regional knowledge about Lake Superior's paleoshorelines and their dates of formation began with such studies as those by Leverett (1929), but a good regional framework was not established until the monumental study by Farrand (1960), who is cited in almost every discussion on the topic. Additional works include Farrand (1969) and an excellent recent summary of this area by Clayton (1985). In addition to higher stands of the lake, the Houghton Stage, about 8,000 years ago, saw a lake surface elevation of 360 feet, almost 240 feet below the present lake surface elevation of 602 feet (Wilson, 1931; Farrand, 1960; Coffin, 1977). If you can imagine a Lake Superior over 400 feet above the lake you now see or a Lake Superior receded to over 200 feet below the shoreline you stand upon now, it is surely evident how natural climate fluctuations were dramatic long before humans were able to be much of a factor.

In summary, the glaciers had a profound effect on the formation of the Apostle Islands. They widened and deepened the stream valleys between the islands when they were simply hills on the mainland. They sculpted some of the bare rock surfaces we see today on the islands, and they brought in the immense amounts of till with its associated clay, sand, and gravel with boulders, which form most of the surficial deposits visible on the islands and on the Bayfield Peninsula. In addition, they laid a thick veneer of glacial deposits upon the bottom of Lake Superior. Even though we see the profound effect of glaciers here in the Apostle Islands, geologically, the interval of this glaciation was brief and recent. It represents only about two thousandths of one percent of the time since the Bayfield Group was deposited.

Geology of the Present in the Apostle Islands

"The present is the key to the past" is an adage that every student who has taken an introductory course in geology has probably heard. This statement, traced to James Hutton, a Scottish geologist who published in the late 1700's, provided one of the concepts that became a cornerstone of reasoning in the geological sciences. The interfaces where solid earth, water, and atmosphere connect are the locations where important geological processes occur. Solid materials are weathered, eroded, transported and deposited, and, here in the Apostle Islands, we can observe surprising changes occurring over a few hours or years. This gives some concept of how such processes must affect the Earth's surface when they act repeatedly through immense spans of geological time. The visitor will see features that result where the processes are dominantly depositional (Fig. 8) or dominantly erosional (Fig. 9).

Within very recent history, we see profound changes. Older residents who grew up in the Bayfield Peninsula recalled wading along beaches that are now under water. In 1986, when we began field work, the water was over a foot above normal; in 1988, the water level had dropped to below normal levels and exposed spectacular potholes carved by glacial streams on the west side of Lighthouse Bay on Sand Island. We had boated right over them in 1986 without even noticing the features. Islands' residents have witnessed whole sand beaches stripped away to bedrock in a few hours by storms, and then saw them rebuilt as longshore currents moved sand back over the exposed rock within a matter of weeks. An abundance of fallen trees and fresh landslide scars along the shoreline confirm rapid erosion of the glacial till. Here along the shoreline, rapid changes occur even in the solid sandstone bedrocks. Arches collapse, stacks and balancing rocks form and then crumble because of the relentless attack of pounding waves.

Mechanisms that cause rapid change in the bedrock shores include ice pushing against low-lying shelves, storm wave displacement of loosened bedrock along the tops of wave-cut platforms (even rocks high above the normal lake level), ice draping from wave spray that adds tremendous weight to bedrock outcrops and apparent pore pressure from springs. Photographs of some of these features taken a few years apart reveal how quickly features change. As recently as 1975, Hermit Island hosted a spectacular arch at its northeastern end, "Lookout Point," which collapsed in the summer of that year. The arch called "Hole-in-the-Wall" on northeastern Oak Island has changed visibly in recent years. We found great variations in the years that we did field work. From our photos taken in 1986 and 1987, for the first edition of this guide, and photos taken ten years later, we were able to document striking changes. We saw major arches at Cat Island and stacks at Stockton Island disappear, and new incipient stacks develop. Some massive low-lying bedrock shelves on inner islands such as York showed changes easily diagnosed as caused by the pressures of winter ice. Sea Cave areas in the Devils Island Formation showed profound changes, particularly along the Mawikwe Bay area. A new set of caves is rapidly developing at the south end of Sand Island. Late winter and spring is the critical time of these events. This is a time when thaw moves water into springs at a time coincident with ice loading on certain cliff faces. Despite the profound changes that we recorded, actual eyewitness accounts of massive slides and rock falls in these islands seem almost nonexistent. Boaters, the most likely witnesses of rockfall events, are not present near these sites in late winter and early spring.

Other changes result gradually. Rock and till are eventually converted to soil, which becomes a haven for living things that likewise undergo life cycles and succession. Perhaps no phenomenon is more natural than change through time. George Collie (1901), one of the first investigators to carefully describe the temporal nature of the Apostle Islands, recognized that "the islands are simply a phase in the history of the lake, their existence or obliteration depending on relatively slight fluctuations of the lake level."

Collie was only one of many in his time to have recognized that the terraces seen on islands such as Madeline Island represented higher stands of the water level of Lake Superior. However, he was unique as one of the earliest workers to recognize that the lake level has been rising in recent times. As evidence, he cited Lake Superior's inundation of stream valleys revealed at the mouths of the streams along the peninsula and also the recent and rapid destruction of sandspits, particularly the sandspit off Grant Point on Madeline Island, and the "Chequamegon Point Bar" (now termed "Long Island" on present day maps). Collie related the rise of the lake level to the statements of Grove Carl Gilbert (1843-1918), who hypothesized that the Lake Superior Basin was actually undergoing tilt and that the northern portion of the Basin was rising relative to the southern. The amount of upward rise is less than 1mm/year near Bayfield, nearly 3mm/year at Outer Island, and greater than 5mm/year in the main body of the lake (Gable and Hatton, 1983). This situation is analogous to drinking from a teacup; as you tilt the teacup to your mouth, the outer edge rises and the edge toward you lowers. Even though the volume of tea remains initially constant in the cup, the tea eventually overflows the lowered edge. The anticipated effect such tectonic tilting will have is slow drowning of local shorelines. Less predictable, however, is the effect of most recent variations of climate and rainfall upon total volume of water in the lakes and therefore upon lake level fluctuations. We know from the general geologic history that water levels have fluctuated widely here; however, we do not know enough detail to evaluate what variations in water levels we should accept as "normal" over a few decades.

Engstrom (1985) summarized the effects that encroaching waters are having on local beaches, spits, and cuspate forelands (see also Engstrom, 1972, 1973, 1974a, 1974b). His measurements over that decade indicate that the beaches on the inner islands are gradually decreasing in size, while a very few on the outer islands are growing. This is because the local sand beaches are thin veneers over the bedrock. The sand is of limited supply and significant amounts come only from the erosion of the glacial till. Beaches or spits occur only where glacial till along shorelines provides a ready supply of sand. However, even there the veneer of sand is usually thin. The supply of sand is limited by the amount of glacial till exposed upcurrent of the beach. When lake levels rise, most spits and beaches in the Apostle Islands shrink.

Beaches, spits, and other depositional forms of sand (Fig. 8) require (a) a source of sand; (b) wave energy sufficient to erode the sand from the shore and to carry it by longshore currents and (c) a platform which is sufficiently shallow and sheltered from dominant waves to allow the sand to accumulate. Those islands that lack a thick cover of glacial till, such as

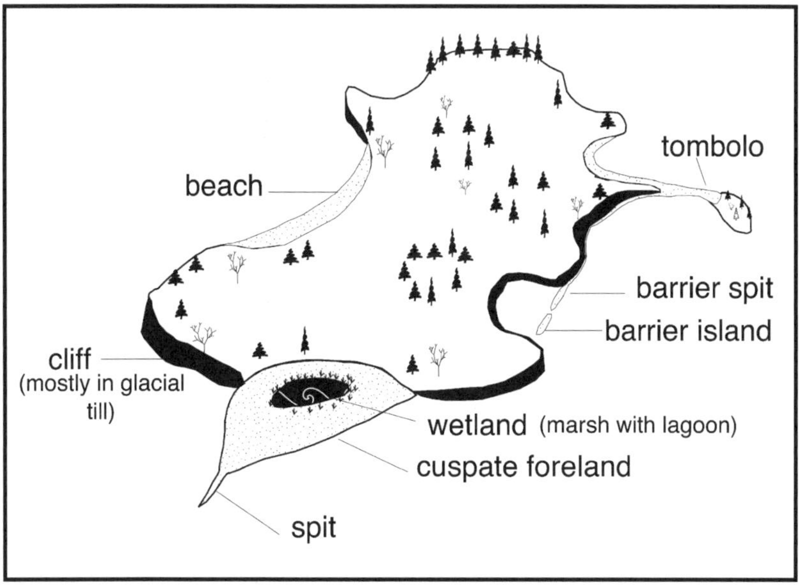

Figure 8. General graphic showing idealized depositional sandforms: tombolo (see also Figs. 59 & 73), spit (see also Fig. 62), beaches, barrier island, and cuspate foreland (see also Fig. 57).

Devils, North Twin, Eagle, and the extreme example of Gull (see Figs. 25 & 26), will lack depositional sandforms even if sheltered areas are present. Those islands with thick glacial bluffs, well exposed to waves that can remove the sand from the bluffs and carry it along shore, will likely have well-developed sandspits on their sides leeward of the dominant wave movement. The sandspits of Outer (Fig. 62), Michigan (Fig. 62), Rocky, Cat (Fig. 62) and South Twin Islands are the best examples.

Rarely, the sandspits may extend all the way to another island, thus joining the two with a sandbar called a tombolo (Fig. 8). Sandy beaches along tombolos include Julian Bay and Presque Isle Bay on Stockton Island and the cove on the northern side of York Island. A cove that is down current of a glacial bluff will almost certainly develop into a sandy beach (Fig. 8). Siskiwit Bay at Cornucopia, Sand Bay on the northwest side of the Bayfield Peninsula, Quarry Bay on Stockton Island, Lighthouse, Justice, and West Bays on Sand Island, and the unnamed eastern bay of Raspberry Island are examples. Sometimes, when sand is very abundant and longshore currents are strong, a barrier island (Fig. 8) is built across a bay or lagoon. An example is at Big Bay State Park on Madeline Island, where a barrier island built across a bay left behind a lagoon isolated from the main lake. Long Island (see Bona, 1990 a, b), which became part of the national lakeshore in 1986, is the only island in the park that does not have a bedrock core. There, sand, brought onto the south shore of Lake Superior by northeastward-flowing streams (such as the Kakagan River and the Bad River) and carried northwestward by longshore currents, has built a long narrow sandspit trending northwestward into the mouth of Chequamegon Bay.

Beaches depend on the stacking of sand onto the shoreline by waves; thus, they are dependent upon wave conditions. Stronger waves, such as occur sporadically in storms or seasonally in winter, will usually have enough energy to move the sands off the beach and pile them up in offshore bars a short distance from shore. When waves become gentler, sand is moved from the bars and back onto the beach. Thus, sand circulates in protected coves. As you land on one of the many beaches in the islands or along the peninsula, you may see that the sand on the upper beaches is segregated into light and dark layers. The dark layers are called lag deposits (see Fig. 55). The black layers are rich in magnetite, an iron oxide, and ilmenite, an iron titanium oxide. These are called "heavy minerals" because they have a higher specific gravity than the lighter-colored quartz and feldspars that comprise most sand. If you thrust a good magnet into these deposits, you will see dark grains of magnetite adhere to the magnet. When storms begin to subside, waves occur with just about the right amount of energy to winnow out the less-dense grains, leaving a concentrate of the black heavy mineral grains (lag deposits) to lie on the beach. If the sand is moved into deep water by storms and settles below wave base, it is permanently lost from the beach.

Although the bedrock is very old, the features sculptured from the bedrock such as arches, sea caves, stacks, etc. (Fig. 9) are products of recent erosion.

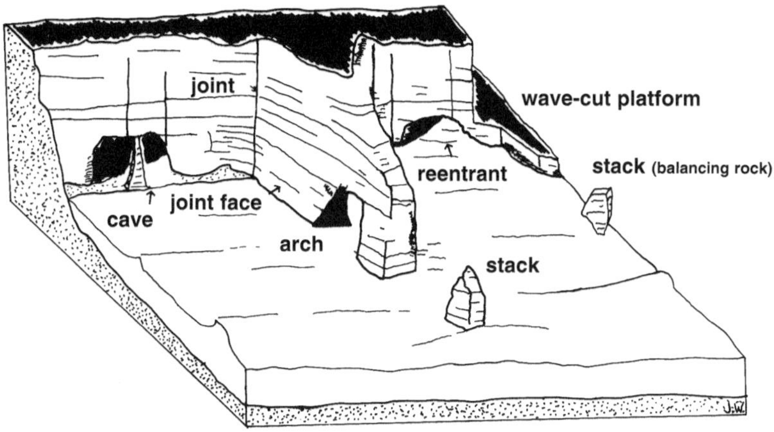

Figure 9. Schematic depicting erosional features common along the rocky coasts of the Apostle Islands: wave-cut platform (see also Fig. 39), arch (see also Figs. 16, 38, & 76), stack (see also Fig. 72), balancing rock (see also Fig. 75), reentrant, and caves (sea caves; note also Figs. 17, 52, 77 & 78). Joints (see also Figs. 64, 72, & 80) play an important part in the formation of arches, stacks, and balancing rocks.

The types of features that form depend upon the bedrock and the exposure to waves. The Orienta and Chequamegon sandstones are dominantly river deposits that are characteristically solidly cemented, thickly bedded (beds greater than two feet), discontinuous, and sometimes separated by less durable shaly beds of sandstone. Where these less durable beds lie near water level, waves erode them and eventually incise back to undercut the more solid overlying beds and form reentrants (see Fig. 9). The reentrant deepens to form an overhang or sea cave, until eventually the overhang collapses of its own weight.

Joints, as previously mentioned, are narrow cracks in the rock; they usually are nearly vertical and occur in groups (sets) that are more or less parallel with respect to compass orientation. Meller (1987) believed that preferential erosion along regional joints could affect the major shoreline orientations of some (Sand, Devils, North Twin and Outer) islands. Certainly, they do control local configuration of some shorelines by serving as zones of weakness that help waves to erode more quickly along the joint faces. Thus, waves wear back the rock unevenly, produce a jagged shoreline with protruding spurs, and undercut reentrants. The northeast corner of Stockton Island is such an example.

Sea caves begin to form when the cutting along the joint spreads laterally to form an undercut reentrant. Where sands are thick bedded, reentrants develop as simple wide cavities along the waterline. When sufficient underlying material wears away, the roof of the reentrant collapses. In order to form spectacular caves, a number of reentrants must join behind the face of the cliff to other nearby reentrants and leave behind supporting pillars and arches (Figs. 9 & 18). This lateral spreading of reentrants can occur most easily where the sand layers that comprise the bed are very thin. Caves are particularly spectacular within the Devils Island Sandstone. There, the sandstones are thinly bedded, much more so than those in the overlying Chequamegon or underlying Orienta sandstones. Where waves erode the thin layers of the Devils Island Sandstone, the thin layers spall off a little at a time, and the reentrant cut by the waves is enlarged gradually and begins to grow into a cave. Erosion proceeds more rapidly along joints, and the reentrants increase in width and in depth. Where cementation is better, the layers persist in place, but the spalling goes on around these until the better-cemented areas (or areas held in place by loading stress) are isolated into columns that remain to support the overlying roof. The erosion of thin layers continues until caves intersect other caves and an entire section of cliff becomes delicately honeycombed with arches and pillars. It is usually obvious where the Devils Island Sandstone outcrops on the shorelines, because this is where the most intricate systems of sea caves are found. Caves on Sand Island, Devils Island, and at Mawikwe Bay Cliffs are all outstanding scenic wonders carved from the Devils Island Formation. In some places, such as the northeastern end of Stockton Island, the Chequamegon Sandstone becomes somewhat thinly bedded, and some intricate sculpture occurs there.

Stacks, pillars of rock isolated from the main island (see Fig. 9), are usually joint-controlled. When waves erode along a shore where a prominent set of joints occurs and where the individual joints are separated by perhaps a few feet of rock, the rock is cut back preferentially along the joints, creating a jagged shoreline. Then the narrow projections of cliffs that have developed between joints begin to break down, and parts of the cliff become isolated. The beds stacked as narrow blocks bordered by closely spaced joints are unstable, and most blocks soon tumble. The cliff recedes quickly, but in a few places the blocks are wider and the stacks of beds more stable. These remain as pillars or stacks long after the cliff has eroded landward, leaving a stretch of open water between stack and island or stack and mainland. The northeastern end of Stockton Island reveals this process rather well. Several stages of stack formation along joints occur at Eagle Island. Sometimes an arch forms in the process of wave erosion, but the center collapses, leaving the stable pillar to become a stack. At the northern tip of Hermit Island is one such collapsed arch.

Wave-cut platforms, flat expanses of rock at or above the waterline (see Fig. 9), are common features, particularly in the Chequamegon and Orienta sandstones. These platforms represent periods of rather constant lake water elevations. Waves, and especially the erosive action of wave-transported cobbles and gravels, form these platforms. Usually they occur where less resistant shaly sandstone lies over more pure, tightly cemented sandstone near lake water elevation. They are particularly characteristic of the eastern shoreline of Outer Island, and of parts of Stockton and North

Twin Islands. Some perched platforms represent older, higher lake levels. Along northeastern Stockton Island, perched several feet above the present lake level, lies an accessible platform that contains sea caves and arches. On Devils Island, the thinly bedded sandstones below the lighthouse are eroded along bedding planes into very broad benches, which display spectacular ripple marks.

The process of shoreline erosion along the glacial bluffs is one of the easiest features to observe. Landslide scars and slumped and fallen trees are abundant along the shores of the Apostle Islands, especially where the red bluffs are most visible. At first it might seem that this erosion within a national park should be controlled, until you realize that this process is the natural one of change through time. The erosion of the bluffs of the islands is the process which supplies sand to maintain the beaches and sandspits that are utilized by visitors for scenic photography, hiking, camping, picnicking, beachcombing, swimming, and sunbathing. Without this erosion to supply sand, the beaches, spits, cuspate forelands, tombolos, and sandy bays for anchorage would soon disappear. The processes of shoreline erosion include erosion by rain sheetwash and runoff, slumping and rock fall, liquefaction and mudflow of clayey deposits, wave action, and pressure from ground water as it pushes against soil and till from springs along cliffs (Edil and Shultz, 1983; Sterrett, 1980; Need and others, 1980). However, the most important of these factors appears to be wave action (Sterrett, 1980). Most erosion occurs in the spring, when meltwaters have soaked the bluffs, and escaping ground water destabilizes them. At that time, spring winds maintain high-energy waves on the lake, which in combination with the shoreline conditions, cause extensive shoreline erosion. During mid-winter, ice around many of the islands does produce some erosion, but mostly the ice serves to protect the shoreline area by keeping the violent winter-storm waves away from the island margins.

As you look carefully at the Precambrian sandstones and compare them with the sands seen along the modern lake, you soon see some similarities. The bifurcating and asymmetric shoaling ripples seen in the Devils Island Sandstone (see Fig. 21) can also be seen forming in the soft sands of bays such as Presque Isle. Current ripples similar to those seen in the Chequamegon Sandstone (Figs. 67 & 69) can be seen at the mouth of Sand River. Cross-beds seen in the sandstones (see Figs. 15, 30, 42, & 54) look like cross-beds seen by digging in river sandbars of today; mud cracks seen in billion-year-old rocks (Fig. 22) look like mud cracks seen on any modern mudflat. The bedded cobbles and sands seen lying on glacial till above the present lake look very much like the cobble and sand beaches of the present, simply because they were formed in the same manner as those you now see. The present is indeed the key to the past—and probably to the future as well.

Geological Tour of the Islands

A number of the islands have features in common. All except Long Island (Fig. 3) have bedrock pedestals as their base. On some islands, such as Michigan Island, the bedrock is covered with Pleistocene glacial deposits, but bedrock lies just below the present lake water level. A few generalizations stated here will help keep you oriented geologically. The braided streams that formed the majority of the rocks in the park (Orienta and Chequamegon formations) flowed from southwest to northeast (see Fig. 5) in the general direction parallel with the long dimension of most of the islands. Figure 4 shows a schematic of a braided stream system. Look carefully at the view you would have depending upon how the stream deposits are sliced. If you cut them at right angles to the flow, you would see mostly cross sections of stream channels (see Figs. 28 & 31). This is the view you would get at the tips of most of the islands, but particularly the northern tips where the rock is better exposed. If you were to cut the deposits parallel to the flow, you would see mostly elongate beds of sand a few feet thick, continuous for perhaps a few hundred feet. This latter is the view you get as you boat along the sides of the islands, particularly the eastern sides of the elongate islands where more bedrock is exposed (see Fig. 41). You will begin to see more and more detail in the rock as you realize the course you are taking along an island or shore either parallels or cuts across the courses of the streams that laid down the sands over a billion years ago. These sands are seen today in the sandstone bedrock.

To see the geology of the area well, you must take to the water. Outcrops on the mainland of the peninsula are small and sparse, and rock exposed in road cuts and abandoned quarries shows little detail. However, boaters can see striking features by proceeding along the shore of the mainland. At the end of this section, we have included two tours along the mainland, which provide ready access to a number of these features. A significant portion of the tour "Bark Bay to Little Sand Point" lies within the national park boundaries. The second, "Washburn to Houghton Point," permits access when west winds create rough water around the islands and around the west and north sides of the Bayfield Peninsula.

Basswood Island

Birth of the Brownstone Industry
Map References—Basswood and Madeline Islands 7 1/2' Quadrangles

Basswood Island is the island closest to Bayfield. The shoreline is steep, and low exposures of the Chequamegon Sandstone bedrock are present but are generally not spectacular. Heavy vegetation on the island prevents a good view of bedrock exposures except at the northern end, where "Honeymoon Rock" exists as a small but attractive stack capped by a thin well-cemented channel sand (Fig. 10).

Figure 10. "Honeymoon Rock," a small stack of Chequamegon Sandstone that has become isolated from Basswood Island. Photos: lower, taken in 1986 and upper, in 1996 show some changes even in the inner sheltered islands. The stack supports a few rugged trees with a miraculous ability to thrive on bare rock under conditions of complete exposure. Lake levels were lower in later photo. Visitors today will see small changes in even a few years.

The brownstone industry, which flourished from about 1870 to the mid-1890's, began on Basswood Island. By the time Thwaites made his study of geology in 1912, brownstone production had essentially ended. The Bass Island Brownstone Company opened the first quarry near the southern end of Basswood Island in 1868 to supply stone for the

Milwaukee Courthouse (Buckley, 1898). Engineers opposed the use of this new and untried building stone because they feared that it might be unsound, but favorable results of tests performed at the Smithsonian Institution in Washington, DC convinced builders to utilize the "Lake Superior Brownstone." The courthouse was built in Milwaukee in 1870, and the quarry supplied stone, mainly to Chicago, but also to localities in Wisconsin, Minnesota, Indiana, Iowa, Nebraska, South Dakota, and Michigan. Products of this quarry were mainly cut stone for buildings and foundations, but an upper layer of broken stone rubble was sold for breakwater cribs and piers. A seven-foot layer of "No. 2 Stone" (stone with a few pebbles and clay pockets) and an underlying ten-foot layer of "No. 1 Stone" (stone without significant flaws or blemishes) were the main seams exploited after removing 4 feet of soil and 8 feet of rock rubble. Blocks shipped were normally 8 feet x 4 feet x 2 feet. A trail leads from the park dock on the west side of Basswood Island to the quarry, which measures about 400 feet long, 300 feet wide and 40 feet deep. A second quarry about a half-mile to the north (Breckenridge Quarry) produced broken stone, but only for a short time.

The stone chosen for quarrying appears to occur along a trend (from Houghton Point through two quarries that lie along Highway 13 at 3.2 and 3.6 miles south of the Bayfield city limits, to Basswood Island, then to Hermit Island, and finally to Stockton Island) that parallels the flow direction of the Precambrian streams (see Fig. 5). Possibly, only a few stream channels provided the right thickness, purity, and color suitable for building stone, and although the beds were not recognized as stream channels at the time, the operators recognized the trend of the occurrence of good stone.

The colorful beauty of many of the structures on the Bayfield Peninsula, particularly the Apostle Islands National Lakeshore Visitor Center in Bayfield, explains in part why the stone was such a desirable building stone. The rich red-brown color comes from a combination in the sandstone of pink feldspar and about 3 percent iron oxide. There are several reasons why the quarries ceased production. Thwaites (1912) cites a combination of demand for a lighter-colored stone in the cities (eventually met by the Bedford Limestone) and the declining reputation of "Lake Superior Brownstone," caused by speculators who quarried and shipped stone of low quality. Buckley (1898) cites "general business depression" as a reason for the demise of the brownstone industry. However, 1880 marked the beginning of the end for quarried stone as the primary architectural medium. That year, the first "skyscraper" came from a new design that relied on steel frames and concrete. This permitted taller buildings, better use of limited building space, and reduced need for quarried stone—especially dark stones. A dark stone such as Lake Superior Brownstone, which is beautiful in an open sunlit area, can produce a dreary appearance in a suite of tall buildings set closely together. These factors combined put an end to the Brownstone industry.

Bear Island

Map Reference—Bear Island 7 1/2' Quadrangle

Kowalski (1976) and Shetron (1974) studied the physical characteristics of Bear Island, and Larsen (1975) and Schultz (1976) described the fauna and flora. The southern part of this island is a highland rising to a hilltop elevation of 835 feet (second in elevation only to Oak Island), but the northern portion of the island exhibits lower and more subdued topography. The southern highland is formed mainly on glacial drift that has been partially reworked by lake waters that existed as higher stands of Lake Superior over the past 10,000 years. The northern half of the island has only thin soil cover over bedrock. The thick drift found on the southern half either was never deposited at the northern end or was eroded away by glacial ice. Lake clays about 11,500 years old were deposited when the surface of the lake (glacial Lake Duluth) was 250 feet higher than at present. These clays dominate the thin surficial deposits on the northwestern end (Kowalski, 1976) and appear deposited on rock devoid of any thick underlying tills. The present topography of Bear Island appears to be mainly a product of older uneven distribution of glacial till.

At time of this writing, the better landing sites are privately owned, so hikers seldom access the interior of Bear Island, particularly the southern part. The higher regions are forested with hardwoods. Old terraces composed of sandy and gravelly material occur at about 810 feet, 715 feet and 685 feet. These are remnants of shorelines formed between 9,600 and 5,500 years ago when lake levels were higher (Kowalski, 1976). Kowalski (1976) also ascribes some sandy soils found along the southern interior of the island as developed on glacial outwash.

A good place to begin a tour is near the large and attractive sandspit (Fig. 11). Several years of data from Engstrom (1985) show that the spit was diminishing in size because of the deepening of lake waters. Lowering of lake levels, more typical of some recent years, produces opposite effects, so sandforms such as spits can vary in size from year to year. As you travel northward along the east side of Bear Island, you first see the steep reddish-brown glacial bluffs. The erosion of these bluffs provides sand for the attractive beaches along the cuspate foreland. Closer to shore, you may observe sandstone boulders and a variety of glacial erratics washed from the till. The first bedrock exposed is a bit farther north, where the channel sands of the Chequamegon Sandstone are visible from the east. This view displays the Precambrian stream channels parallel to their general flow direction (Figs. 4, 5). Recent rockfalls result from the erosion of more easily eroded shaly sandstone to leave overhangs above reentrants. A close view of these rocks reveals northeastward-dipping cross-beds, a dip direction that reveals the northeastward flowage of the streams that deposited the sandstones over one billion years ago. Farther up the coast you can see large channel bars with planar cross-beds and some smaller scour cross-beds in the sandstones. As you enter along shallow coves toward the northern part of the island, you can find some low but intricate sea caves and reentrants.

Figure 11. Sandspit developed along the cuspate foreland (see also Fig. 57) at the southeast tip of Bear Island. Till in bluff in background at right serves as a source of sand. The foreland is large and contains a well-established stand of trees as well as grasses and a sandy beach. The beach serves as a popular site for gulls and other water birds.

The northern rocky portion of the eastern side of the island is worth landing on and exploring, where a marvelous exposure shows sand bars from the billion-year old streams in detail in three dimensions. The outcrop shown in figure 12 reveals rapid deposition of a sandbar built over the gravel-strewn stream surface. The planar cross-beds dip in the direction of flow of the Precambrian stream. The rock outcrops between the narrow bays show a variety of cross-beds and conglomerate (coarse rounded gravel) zones extensively exposed (Fig. 12).

In some areas of the conglomerate, small potholes are forming. Most potholes form in swift streams below rapids and waterfalls, but here, wave action apparently provides sufficient energy to form small potholes. Glacial erratics (rocks from outside the region brought here by glaciers) of granite, gneiss, and metabasalt serve as tools because they are more durable than the local sandstone. These enter a depression (left by a pebble or cobble washed from the conglomerate) and then wear the depression ever deeper and wider as waves move the tool back and forth and around in the depression. Eventually, a smooth pit or pothole several inches deep is carved (see Fig. 56). Several potholes with the hard tools still in place may be found here. Lifting the tool from the pothole reveals the smooth polished surface produced on both tool and pothole in the grinding process. (If you do remove a tool to examine it, please make sure to return it to its original position so that the process can continue and others may view it.)

Figure 12. Photo taken in 1986 reveals a conglomeratic sandstone exposure at northern end of Bear Island. Note the flat area in the foreground. It is an actual surface of an old streambed. The inclined beds are cross-beds where an overlying sandbar was progressively built across the streambed surface. Cross-bed inclinations reveal stream flow from left to right (SW to NE). Tangential bases of the sandbar's cross-beds rest in megaripple troughs on the conglomeratic bed. Thin conglomerates, some oriented in patches transverse to stream flow, are noted in modern braided streams.

Much of the rocky shore along the northeastern part of the island consists of low, rocky flats of wave-cut benches presently being formed. Some small waterfalls spill onto these rocks as streams drain from the island. The sands found here on the tiny northern beaches are different from any sand found elsewhere on spits and beaches in the park. They are very coarse, over 90 percent quartz, and rich in polycrystalline quartz. Because of the abundant polycrystalline quartz, Nuhfer and Hodgson (1987, 1988) postulated that this particular sand might owe more of its source to the local bedrock than do other sands whose source is glacial till. A particularly good place to see this peculiar sand is where the major stream leaves the island just south of a small cabin. Strzok (1981) has noted that small waves on the lake will build here to yield surges about twice the height of lake waves as they squeeze through the ledges bordering the coves. These surges are likely responsible for winnowing finer grains and causing concentrations of the notably coarse sand here.

Toward the northerly portion of the island, joints may be observed on the benches and in the clear water just offshore. Several prominent ones that trend northeast toward Rocky Island may be seen. The combination of jointing and the exposure to waves from northeasterly storms creates a rugged shoreline with attractive cliffs, grottos, caves, and reentrants.

Glaciers were doubtless responsible for peeling away upper layers of rock along the northern edges of the islands. Some glacial grooves may be seen, but most are on broken slabs. A few may be seen in place on the most northerly promontory of Bear Island. As you round the tip toward the westerly side of the island, a few small sea caves and arches come into view. These rocks also display many cross-beds and channel cuts. (Recall Fig. 4 and remember that, as you round promontories and island tips and look in a southwesterly direction, you are looking back in the direction from which the Precambrian streams flowed, rather than viewing them from the side, as you do when you observe them looking directly to the west. Therefore, you will see more cross-sections of channel cuts when looking to the southwest.) Southward along the west side of Bear Island, the rock is draped with glacial till. Glacial bluffs rise higher as you proceed toward the south end of the island and, eventually, eroded gullies become almost canyons. Glacial erratic cobbles occur in the bluffs and accumulate along the shoreline. The erosion of these bluffs furnishes a good supply of sand to the cuspate foreland to the east. Cobbles give way to sand as you again approach the cuspate foreland from which you started.

Cat Island

Map References—Stockton Island and Cat Island 7 1/2' Quadrangles

Cat Island is presently one of the less visited islands in the park, but geologically and recreationally, the island offers many attractions. Most visitors will probably approach Cat Island after departing from Stockton Island and probably will encounter the sandspit at the southern end first. The sandspit (found on the Stockton Island Quadrangle) is fed mainly through weathering of glacial till bluffs along the west side and to a lesser extent from the

Figure 13. West edge of sandspit at southern tip of Cat Island viewed looking south.

east. Engstrom (1985) notes that this sandspit (Fig. 62) is diminishing somewhat on the west side, but growing on the east. The spit maintains a zoned growth of forest, grasses, and a rim of sandy beach veneered with small driftwood (Fig. 13). Much gravel is found with this sand and in places gravel bars are present, their extent depending upon recent weather. Gentler waves associated with calmer weather tend to pile sand over the gravels.

Occasionally, you'll notice that when you walk on dry sand here, the impact of your footprint produces a series of tiny collapse pits in an area a foot or two around each footstep (Fig. 14). It results from gas bubbles trapped in wet sand. When the sand dries, cohesion between grains decreases, and then even the vibration caused by a footstep can produce collapse into the old cavities left by bubbles. This phenomenon is not restricted to Cat Island. It can be observed on many beaches.

As you proceed to the east side of the spit, the wide beach disappears, giving way to a shoreline with narrow sand strips, underbrush, slide debris and boulders. Farther north, the first low cliffs of Chequamegon Sandstone are seen. These have a few low wave-carved reentrants with small pillars. A few outcrops reveal some steeply-inclined planar cross-beds (Fig. 15). Such cross-bedded sandbars formed in the streams and "migrated" downstream because of sand being moved from the upstream side, carried across the bar, and deposited on the downstream side. Where locally abundant, these bars could be considered as very large ripple forms sculpted in the soft sands of the channel. They are sometimes called "megaripple" or "giant ripple" beds, depending on whether their sizes are a few meters long or a few tens of meters respectively (Reineck and Singh, 1980).

Figure 14. Pits produced from collapse of bubble cavities below the surface of sand on sandspit at Cat Island. Slow walking on some dry sands found close to the water level will produce about a two-foot diameter ring of collapse structures around each footstep. The cavities result from trapped air and gas formed when decomposition of organic material created gas bubbles beneath the surface of wet sand.

Figure 15. Inclined planar foreset cross-beds on Cat Island. These were formed on the edges of migrating sandbars in the channels of the Precambrian rivers. Each cross-bed represents a layer of sand, which was deposited over the downstream edge of the bar until a bed of sand composed of inclined foreset beds was formed. After burial, this sand was lithified into sandstone with the cross-beds still preserved.

To the north of these low cliffs, a wave-cut platform extending along the shore for about 100 yards is littered with boulders and cobbles of Chequamegon Sandstone. This platform gives way to another set of low cliffs that exhibit some convoluted bedding near the waterline. These convoluted beds formed when small sections of soft sediment in the old rivers slumped under the force of gravity to produce local folding of the cross-beds (see Fig. 44). Several precariously rooted trees cap the cliffs. Some tiny sand beaches lie to the north of these cliffs, and at the north end of the beach a sandstone shelf extends some distance out into the water. The regional joints well exhibited on this shelf as long straight fractures in the rock occur in sets that trend either northeast-southwest or nearly east-west. Some outcropping slabs reveal coarse conglomerate beds and scour cross-beds.

Farther, up the coast, some thin shaly beds of sandstone occur in low cliffs. Where the rock is very shaly, it acts as a sealing layer. Ground water percolates down to this layer and then flows out as springs along the shoreline. The movement of ground water causes leaching out of the iron pigment, and here, as in many places on the islands, the sandstones often show mottles of white within otherwise red-brown sandstones. The white zone here along the line of springs shows the association between color leaching and ground water. Old stream channels cut into the shaly beds here, as revealed by discontinuous sandstone beds and inclined cross-beds lying upon the flat shaly beds. Few, if any, true shales exist in outcrops; most shaly zones have greater than 50 percent sand grains and are thus more correctly termed "shaly sandstones." The shaly deposits probably represent overbank or floodplain deposits along the old rivers. They are more easily eroded than the more massive sands deposited in the main channels; this probably accounts for their relative scarcity in the Apostle Islands. Possibly, they were more abundant in those areas that are now the deep channels between the islands. Where shaly zones occur, they are usually as discontinuous pockets between channel sandstones.

Continuing northward, the cliffs become higher and more prominent. Just south of the sandy bay are a few rocky coves, where the channel sands may be viewed in three dimensions. If you look at the rocks along the upper portions of these cliffs, you will notice that some are brecciated (severely broken and fractured) and appear as unstable rubble. This brecciation of the sandstones is attributed to the immense pressure of overriding ice during the last glaciation. A few boulders here exhibit wide but shallow and flattened erosional pockets (see, for example, Fig. 70). These may be where clay galls or shale pebbles were weathered from the rock. Where sticky clay is especially abundant, it will sometimes roll into a ball rather than be simply dispersed in suspension. When such lumps enter a stream, they may be quickly rounded within a short distance of transport. Streams or waves carry these balls or "galls" just as they do any other gravel-sized particle. Eventually they become buried in channel and bar sands. Compaction flattens them (note the flat shallow shape of the pockets in the boulders) as the soft clay or shale is deformed plastically under the weight of the overlying sediment. After the rock is lithified and exposed later by erosion, these clay or shale galls weather out easily and leave behind the characteristically flat erosional pockets.

The sandy cove along the east side of the lowest and narrowest portion of the island affords some shelter. The beach is narrow, and the cove near shore is littered with boulders of basalts, granites, and other glacial erratics eroded from glacial till. Farther offshore, the larger boulders are probably local Chequamegon Sandstone.

To the north of the cove, cliffs again appear as well as low but attractive reentrants, arches and sea caves. Erosion of these has exposed the rounded bases of channel sands, each channel being only about two feet thick. In 1986, when we approached the northeastern edge of the island, we encountered what we termed "arguably the most spectacular wave-cut arch in the islands" (Fig. 16 lower). We also wrote that: "This arch in itself is worth the trip to Cat Island." Unfortunately, visitors seeking it from the older guide won't find it. By 1996, the arch had collapsed and evidence for its presence was swept away (Fig. 16 upper). Of course, the same wave processes that remove arches are still at work in the islands making others, but the loss of the arch on Cat Island shows how quickly changes occur here.

As you pass around the northern end of the island, you may enjoy some views of low, eroded sandstone grottos and low caves with delicate pillars. The final stages of development of a stack may be seen here where a small promontory of sandstone is yet delicately connected to the mainland. The foreset beds of these channel sands dip steeply to the northeast,

Cat

"What kinda cat is this Cat?"
"Purty as a cat's eye, that Cat,
and as good a catercousin
as you'll ever want."

"Let the Cat out of the bag?
Why, it'd take another St. Croix Advance
to do that, wouldn't it?"

"Feelin' catty?
Nothin' Cat cain't cure!
You'll feel like yer damn near
sittin' in the catbird seat
on Cat."

"Don't tell me I been catfootin' around;
I spent my weekend on Cat."

"When I start catalogin' my blessin's,
top of the list is–Cat."

"Now don't go caterwaulin' around
for Cat.
Outer's where we're goin' today."

"Just a catspaw, Lord;
anythin' more and we'll be stranded
on Cat."

"Let's put a stop to these Cat and mouse games."

Figure 16. Photos: lower, former arch near northeast tip of Cat Island in 1986. The top of the arch that bridged to the island consisted of a massive channel sand pedestal that rested on a column of thinner sandstone beds through which an expansive window was cut by waves. Note the supporting soil cover and trees. Upper, site of the arch in 1996. All that remains is the pedestal. Changes such as this are punctuated events, not gradual wearing away.

in the direction of North Twin Island. A short distance to the west, glacial till covers all bedrock, with the last outcrops being so heavily stained by glacial till as to be inconspicuous. The west side of the island is characterized by bluffs of glacial material that rise from 12 to 60 feet above the waterline, with the exception of a cove on the west side of the narrowest (also lowest) part of the island, which contains a sand beach. The bluffs are highest toward the southern tip of the island and in places are dissected by gullies. A sand beach is encountered as you approach the west side of the attractive sandspit.

Devils Island

The Island of Sea Caves

Map Reference—Rocky Island 7 1/2' Quadrangle

This tiny island (318 acres) is the crown jewel attraction for the tour boats and provides memories that perhaps last after all others fade. A tour of the coastline in a kayak, small boat or dinghy offers a truly outstanding adventure. The geological features of greatest attraction here are sea caves produced by wave erosion (Figs. 17, 18, 19 & 20) and the billion-year-old ripple marks (Fig. 21). The island serves as the type locality for the Devils Island Formation, which in fact occurs only on the northern two-thirds of the island (Fig. 3). The southern portion of the island near the boat slip is composed of the lowermost beds of the overlying Chequamegon Sandstone. The Devils Island Formation (see section on general geology) was deposited on sand flats where shallow streams entered a lake. In such an environment, sands deposited are thinly bedded, fine-grained, and extensively ripple-marked.

Figure 17. Intricate sea caves formed in the Devils Island Sandstone at Devils Island. Caves occur on both the east and west sides of the northern half of the island.

Figure 18. Pillars and caves on Devils Island. The pillars exhibit thin bedding. Waves continually enlarge openings into major caves.

The shape and orientation of Devils Island will generally yield calm waters on either the east or west side in the immediate vicinity of the caves. For ideal lighting, the east side should be traversed in morning and the west side in early evening. There are few scenes anywhere that can rival the Devils Island caves near sunset. They are not ordinary reentrants but are delicate rock sculptures—lacelike in their detail, full of brilliant colors, alive with sound and with the motion of surging water. Characteristic sounds from the sea caves may often be heard at great distances.

When the lake is rough, the sounds are actually explosive booms coming deep from within the cave. Even in calm weather, when the caves may be entered with small boats, the sound has a genuine quality of percussion or impact caused mostly by air and water acting together on the rock. In the deep recesses of the cave, it is possible for air to be trapped as an incoming wave seals the gallery from top to bottom. The momentum of the wave generates a continued push into the cave, and water compresses the air, much as a piston does in the cylinder of your car engine. A deep recess may have air compressed by an extensive mass of water. However, as the wave recedes, the water is pulled quickly away, and the compressed air rushes outwards with explosive force. It is easy to see how a storm with accompanying increased height, speed, and frequency of waves could generate powerful forces at the back of caves—forces capable of eroding solid rock and of generating explosions that can be heard for long distances. Most certainly, one can see many changes from year to year on Devils Island if one compares earlier photographs of the same portions of the shoreline (Figs 19, 20).

The lighthouse keeper at Devils Island reported to us that, in places, the caves have contacted fissures that reach the ground surface. He reported that during unusually violent storms, water and air shoot out of

Figure 19. Ten years of change in erosion of Devils Island Sandstone. Photos: lower, taken in 1986 and upper, taken in 1996. Development manifests in many places as enlarged caves and removed slabs.

Figure 20. Ten years of effects of wind and waves on Devils Island. Note large pine tree in center (left, 1986) that had been snapped away at trunk by 1996 (right) and general differential erosion of softer beds over the decade.

these fissures like geysers, apparently driven by the force of large waves compressing air as they rush into the caves below.

As discussed in the section on general geology, the thin-bedded character of the Devils Island Formation is responsible for the development of these intricate sea caves. The result of waves acting on these thin-bedded sandstones is to produce erosion in several directions so that the caverns are enlarged ahead, to the sides, and upward. Pieces of the thin layers are removed, but areas that are more cemented, slightly better positioned, or supported against impact remain. It is evident from observing these caves closely that they were not necessarily formed along a particular bed of weakness nor did they develop simply from the waterline upward.

Why do pillars (Fig 18) exist in certain areas totally exposed to the elements even though they are not composed of rock visibly different from the surrounding rock? These columns of thin-bedded sandstones are apparently stabilized by the overlying weight they support—much as a deck of cards supported by a weight remains in place against wind. The lack of any standing isolated columns without roof contact appears to indicate that the columns may owe their existence to compressive stress directed from above into selected areas rather than to any distinctive property in the rock forming the column.

Variable bedding is another factor governing cave development. Often the existence of a thicker channel sand is seen to control the height to which the cavern will develop and perhaps even to govern the temporal existence of the cave itself.

Although the ripple marks may be seen in slabs along the water's edge and in the sea caves themselves, the best views are provided along the cliffs below the Devils Island lighthouse on the northern tip of the island (Fig. 21). Erosion has exposed extensive surfaces where you may see layer upon layer of rippled sandstones—remnants of a few days' activities on the sands over one billion years ago. They are colorful and spectacular, worth seeing even if water conditions dictate a hiking trip from the boat slip at the southern end of the island.

Observers have mistaken some of the ripples on Devils Island as "current ripples" because of their asymmetric profiles (one side steeper than another). However, the bifurcations and straight-crested nature (Fig. 21) reveal that these were produced by shoaling waves rather than by currents. The same type of ripples can be seen forming today where shoaling waves sweep toward the beach and across the sands in Presque Isle Bay on Stockton Island. Other ripples do exhibit the symmetrical sinusoidal profile typical of a classic ripple produced by oscillating waves. A few true current ripples are also present here; these were probably produced by streams meandering in almost random directions over the sand flat. These streams produced channels with associated scour troughs and cross-beds. However, the northeasterly flow direction clearly evident in the overlying Chequamegon and underlying Orienta sandstones is not at all evident here. Further evidence of the emergent nature of the area during Precambrian time includes mud cracks (Fig. 22) and vortex ripples (ripples in small doubly-tapered patches associated with back and forth movement of water on irregular, shallow surfaces—see poem backdrop, p. 93 and Allen, 1982).

Figure 21. Ripple marks at Devils Island. The particular ripple marks found here are mostly shoaling ripples formed by small waves moving (from left to right in this photo) toward shore. They often have fork-like branches and are formally termed "bifurcating ripples." These form only in shallow waters with an oscillating wave-like motion, never in uni-directional currents such as are produced by streams and rivers.

Figure 22. Mudcracks in the Devils Island Formation. Both mudcracks (molds) and their fillings (casts) may be seen, together with the ripple marks, below the lighthouse.

Eagle Island (Restricted Access)

Map Reference—Squaw Bay 7 1/2' Quadrangle

Eagle Island is the westernmost island of the Apostle Islands Archipelago. It is small (less than 30 acres) and is the only island underlain entirely by the Orienta Formation. Access and approach are restricted because Eagle Island is a nesting site for several species of birds, and visitors should not approach any closer than within 500 feet from shore between May 15 to September 1. If you wish to plan a trip to this island, check first with the Apostle Islands National Lakeshore Visitor Center. Because of its small size and open exposure, Eagle Island can be a particularly inhospitable little rock. A small boat can be trapped there with little shelter, and weather conditions change quickly in this section of the lake. As noted by Strzok (1981), hidden rocks, especially along the south end, make approach to the island dangerous.

A few interesting features can be viewed from the water while drifting off the western shore. If possible, motors should be shut off at some distance up current of the island so as not to frighten the water birds that you will enjoy seeing as you drift or row quietly past the island. A good set of binoculars will enable you to see the geological features and the water birds, and yet to remain at an acceptable distance from the island. The character of the bedrock is much like the exposures of the Orienta Sandstone that can be more easily examined near the lighthouse on Sand Island. They are light-colored massive sandstones that exhibit scour cross-beds deposited by north-

Figure 23. Parting lineation at Eagle Island. The oriented spalling is thought to be caused by preferred orientation of longer dimensions of sand grains and uneven segregation of fine and coarse-grained sands into strips parallel to current flow.

Figure 24. Newly completed stack formation seen in 1996. North is to the left. Erosion along a joint has severed the southern part of Eagle Island from the main island.

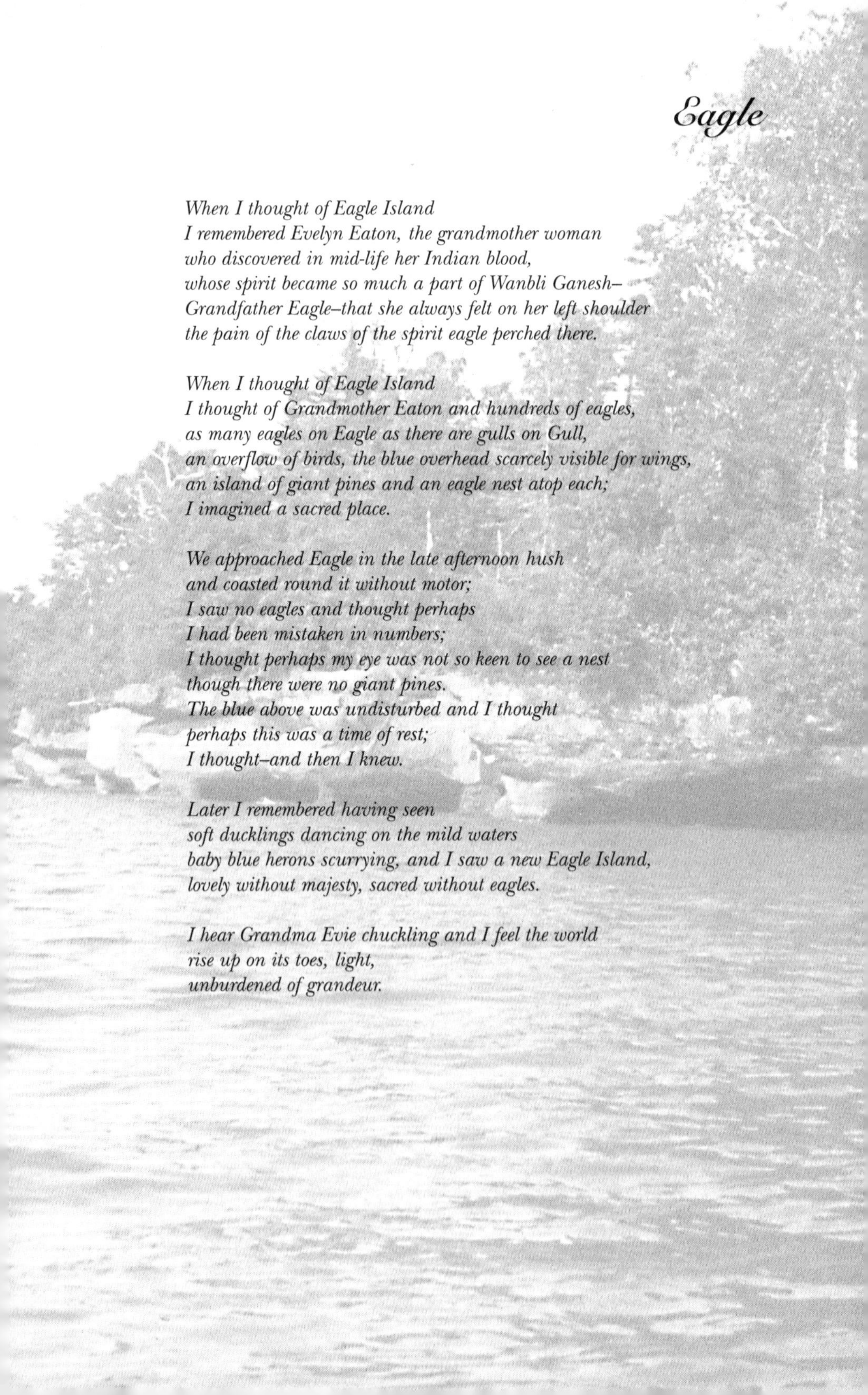

Eagle

When I thought of Eagle Island
I remembered Evelyn Eaton, the grandmother woman
who discovered in mid-life her Indian blood,
whose spirit became so much a part of Wanbli Ganesh–
Grandfather Eagle–that she always felt on her left shoulder
the pain of the claws of the spirit eagle perched there.

When I thought of Eagle Island
I thought of Grandmother Eaton and hundreds of eagles,
as many eagles on Eagle as there are gulls on Gull,
an overflow of birds, the blue overhead scarcely visible for wings,
an island of giant pines and an eagle nest atop each;
I imagined a sacred place.

We approached Eagle in the late afternoon hush
and coasted round it without motor;
I saw no eagles and thought perhaps
I had been mistaken in numbers;
I thought perhaps my eye was not so keen to see a nest
though there were no giant pines.
The blue above was undisturbed and I thought
perhaps this was a time of rest;
I thought–and then I knew.

Later I remembered having seen
soft ducklings dancing on the mild waters
baby blue herons scurrying, and I saw a new Eagle Island,
lovely without majesty, sacred without eagles.

I hear Grandma Evie chuckling and I feel the world
rise up on its toes, light,
unburdened of grandeur.

eastward-flowing streams. The undulating tops of the sandstone beds reflect the undulations of the sand megaripples and giant ripples in the channels of the Precambrian rivers. A nice exposure of parting lineation is present (Fig. 23), and some good glacial grooves are present. However, the best grooves are on broken slabs rather than in place.

The southwestern end of the island displays several stages in the ongoing development of stacks, which are formed by weathering along joints. The set of joints parallel to the shoreline has provided smooth planar surfaces that expose cross-beds and megaripples and serve as zones of weakness where increased erosion has occurred. Another set of nearly east-west joints cuts the rock into blocks. A stack has been newly formed by isolation of the rock along two joint sets near the southwestern end of the island (Fig. 24) where the waters working along an east-west joint have just severed the rock from the main island.

About 40 feet to the north of the site of figure 24, erosion may be seen along another east-west joint, but it has not developed to such an advanced stage. Thus, in a short area about 200 feet long, a systematic sequence of the stages of formation of stacks may be seen.

Gull Island (Restricted Access)

Map Reference—Michigan Island 7 1/2' Quadrangle

Gull Island is a tiny three-acre island found off the northeastern tip of Michigan Island. It was undoubtedly a part of Michigan Island, and a continuous rock reef extends between the two islands at a water depth of less than two feet. Water often rushes through this reef in strong currents and makes control of small boats difficult. Rocks surrounding the island make approach by larger boats very dangerous. Although the island is small, it reveals the nature of islands without glacial drift. No sand beaches are present because the only glacial material that could serve as a sand source is a thin veneer at the western end (Fig. 25). Gull Island's appearance is much like what many of the islands would be if glacial debris had never been deposited. It also represents the stage to which some of the islands will evolve in a few thousand years as the persistent action of waves strips away the last remnants of glacial drift and soil.

Like Eagle Island, this island serves as a nesting site for birds, especially gulls, and visitors should not approach any closer than within 500 feet from shore between May 15 and September 1. Because access is restricted, visitors who plan to be in the area of Gull Island should first check with the National Park Service to learn about current rules for approach. There is probably no good reason to land on the island. It's good to view birds from a distance, but the island itself is most unpleasant for people. Most of the island is low gravel beach (Fig. 26), which is comprised of Chequamegon Sandstone gravels and cobbles. The island appears white from a distance, not because of the color of the rock, but because the brown rocks are covered with a white coating of guano (bird excrement) that testifies to the popularity the island holds for birds.

Figure 25. Thin remnant of glacial material on southeastern side of Gull Island in 1986. The glacial material is also the only soil source, a fact that explains why the only vegetation lies at one end, where the thin remaining soil supports some tangled trees and shrubs.

Figure 26. Gravel beach of Gull Island viewed from north end in 1986. The white color of the otherwise brown sandstone rocks comes from a complete coating of guano (gull excrement), which testifies to the population of water birds and their affection for Gull Island. For this reason, the island is unpleasant to visit.

Gulls

The gulls of the Apostle Islands
dizzy with freedom,
a world of sky held within their wings,
launch a flotilla of human dreams,
inhabit the vision of a thing unstained,
the outreaches of desire.
Wanton wanderers,
winging upward from the mouth of water
to dance in the sky,
light a wild restlessness within the prudent eye,
mediate expansiveness for the timid mind.
But do you know that the gulls' home in the Islands,
Gull Island,
is a three-acre heap of rock
streaked white with guano,
just a handful of soil at one end–
and a swatch of grass–
clutching the stones
to prevent being washed away,
that the nests of the gulls,
fashioned only of heavy gray sticks, nothing soft,
lie on the rocks and the handful of sand,
painfully earthbound,
that the breezes of Gull Island
reek of guano,
that Gull Island is bounded by
waters putrid with guano,
cobbles slimy with algae and guano.

Hermit Island

Map Reference—Oak Island 7 1/2' Quadrangle

Hermit Island is underlain entirely by the Chequamegon Sandstone. It is symmetrically shaped and rises to a maximum elevation of 778 feet near the center of the island. The steeper slope between elevation 680 and 720 is the approximate site of an abandoned beach that appears on the regional map of Clayton (1985). The best exposures of rock lie near the eastern edge, which is easily approached by small boats. The major tour

Figure 27. "Lookout Point" on eastern end of Hermit Island. These testify to the sculpting force of the lake waters, even along the coastline of relatively sheltered Hermit Island. Photos: lower, taken in 1986 and upper, taken in 1996.

boats often cruise a very short distance from Hermit Island and provide a good view of its cliffs and quarry.

The northeastern end reveals a number of joints exposed in cliffs of shaly sandstones and channel sandstones. In places, the joints are so closely spaced that the thin-bedded units appear broken into many blocks, thus giving them a nearly brecciated appearance. Many fallen blocks from rockfalls are visible. As you approach the southerly portion of the eastern end, you encounter a striking promontory built up of channel sandstones, with thinner channels at the base and a massive channel near the top. This is the remnant pillar of an arch ("Lookout Point" Fig. 27), which collapsed in the summer of 1975 (Strzok, 1981). A comparison of the photo on page 29 of Strzok (1981) with our figure taken in 1986 revealed the swiftness of the geological processes of change along the rocky cliffs of the Apostle Islands. Within five years, very discernible differences in the shape and positions of even massive, resistant rocks could be seen.

The nearby cliffs to the south of "Lookout Point" are scenic nesting places for gulls; in late spring, young gulls may be seen among the ledges. The ledges are built along thick channel sandstones that in places are interbedded by shaly units. Some channels contain gravel and have pebble accumulations near their bases. Large slabs and boulders protruding from below the water reveal that large slides and rockfalls have occurred here. Some of the slabs exhibit shallow weathered pockets left from erosion of clay galls. A particularly striking channel exposure is seen a bit farther south and west along the coast (Fig. 28).

Figure 28. Channel-fill sand on Hermit Island. This exposure, now a bit more overgrown since this 1986 photo, still shows a large channel-fill in the Chequamegon Sandstone. The lower rounded part of the channel bottom is exposed by erosion of the underlying shaley sandstone.

Hermit

Renegade Wilson, maker of wooden tubs,
lost a day-long fistfight
for kingship of Madeline
and, banished, came to you.
Living in isolation for ten years,
he gave you your name–
Wilson's Island, Hermit Island,
Hermit.
You could have been named Treasure-trove
for the $8000 in gold and silver
some say is hidden here,
or Prentice's Island
for the owner of the brownstone quarry,
or Heartbreak Honeymooners' Island
for Prentice's new young wife
who wouldn't spend even one night here,
or Eden Island for the apple and pear orchard
above the southeast cliffs.
But you are Hermit, after cooper Wilson,
and forever colored by the name.

How were you to him, little hermit?
Being second choice,
were you less than generous,
making him pay for preferring Madeline,
or did you realize he couldn't live
anywhere but among the Islands–
and make allowances for that?
Did he swear at you, curse you,
and did you hide the sounds of his curses
with sweet rains and wild winds?
Did you fill his bruised and swollen hands
that first week with ripe berries?
Did you take him to your heart
and win his heart?
Did the renegade prefer you in the end?
Little gray hermit, are you Hermit Island
because you so embraced the tubmaker
that your hearts are one?

As you proceed farther west along the southern shore, the shoreline becomes covered intermittently with glacial drift and vegetation until you encounter the site of the old Excelsior Brownstone Company quarry. Excelsior was a subsidiary company of the better-known Prentice Brownstone Company. The site on Hermit Island is easily visible from the water; large blocks of already quarried dimension stone lie stacked and ready for shipment. This quarry was set in operation in 1891, with the first shipment of brownstone sent to Buffalo, New York. The quarry produced at a capacity of 200,000 cubic feet per year. The Chamber of Commerce in Detroit, MI, and the Masonic Temple in Ashland, WI, were built from stone taken from here (Buckley, 1898).

A narrow sand beach exists in the shallow cove farther to the west. Most of the remainder of the shoreline is covered by glacial drift over a few poorly exposed outcrops of Chequamegon Sandstone. Another small sand beach exists on the north side of the island at the mouth of a small stream.

Ironwood Island

Map Reference—Cat Island and Stockton Island 7 1/2' Quadrangles

Ironwood Island is a nearly circular island and therefore lacks the NE-SW elongation of most of the other islands. The most prominent rock outcrops are on the northeast. The western side is covered by heavily vegetated glacial drift.

The sandspit on the south end of the island is attractive. It has been studied by Middleton (1983) and Engstrom (1985). Their work revealed that the spit had grown slightly during recent years through the mid-80's. A logging camp was present on the sandspit, and mounds seen along it were probably artificially constructed when bulldozers buried refuse in the late 1950's. Many large glacial boulders lie beneath the waters near the sandspit, so care must be taken when approaching and leaving.

As you proceed up the eastern side of the island from the spit, channel sands of the Chequamegon Formation are seen in low cliffs. Ironwood Island lacks spectacular standing rocks, high cliffs or arches, though it does have a few small shelves or wave-cut platforms. The sandstones are discontinuous, and individual beds thicken and thin over short distances. Few thick channels are present. Waves have eroded certain beds selectively near the water line to form low arch-shaped reentrants. Several exposed cliffs show convoluted bedding formed during the slumping of the old Precambrian sandbars when they were soft sediments about one billion years ago. A number of recent rockfalls have strewn boulders along the shoreline. Evidence of prominent erosion of bedrock is common. The cross-beds present dip northeastward in the direction of Cat Island. Some cross-beds toward the northeast end of Ironwood Island are thick foreset beds. Scour-and-fill cross-beds are abundant, occurring in many places along the shoreline. These are associated with a complex array of small channel sands.

Long Island

Map References—Long Island and Chequamegon Point 7 1/2' Quadrangles

Long Island became a part of the Apostle Islands National Lakeshore in 1986 and is (Figs. 3 and 29) unlike any of the other Apostle Islands. The most detailed geological study to date was made by Lisa Bona, in a Masters thesis done at University of Wisconsin–Madison (Bona, 1990b). It is a young island, certainly less than a few thousand years and likely even less than a thousand years old (Bona, 1990a; 1990b). It is composed of sand largely derived from glacial deposits, washed into the bay by streams and carried here by northwest-trending longshore currents. There is no obvious shallow bedrock pedestal beneath Long Island. Bedrock is probably not far beneath the sand, although it proved below the reach of hand coring devices used by Bona that could sample only down about ten to twelve feet deep. The island is the distal part of an extensive barrier spit (terminology of Leatherman, 1982) about eight miles long and mostly less than 100 yards wide, which has been slowly built northwestwards into Chequamegon Bay. Although waves and currents continually erode the sand, the spit persists because abundant sand supplied by longshore drift exceeds the volumes that are eroded (Bona, 1990b).

The natural history of the formation of Long Island was first deduced by Collie (1901). Oak Point represents one of the earlier developmental stages; it began as an offshore bar, but eventually, with the addition of sand carried from the mouth of Bad River, it grew to a barrier spit joined to the

Figure 29. View looking south across eastern Long Island. Low dunes of white sand and the sandy beach ridge behind are being stabilized by grasses. The low area between, covered with pebbles and larger driftwood, is a channel carved by storm washover.

mainland. After this barrier spit was established, an extensive wetlands formed within a bay behind the southern edge of the spit. As more sediment was added, the marsh occupied more of the former bay. Longshore currents continued to carry sand past Oak Point and accumulated in a point bar, termed by Collie (1901) as the "Chequamegon Point Bar." This point bar grew into a long barrier spit, of which Long Island represents later stages of building.

It is easy to document the dynamic nature of this island from the literature. "Long Island, the remarkable island four miles in length which forms a part of the natural breakwater at the entrance to Chequamegon Bay, is an island no longer, but simply a very long and narrow peninsula, extending out from the mainland. No one can guarantee its present status very long, for within the memory of the present generation at least, Long Island has been quite fickle" (Burnham, 1930). Recent navigation charts as well as the current (1963; field checked 1964) Long Island 7 1/2' Quadrangle map show Long Island separated from the main peninsula by a wide channel. However, during the field season of 1987, Long Island was again a continuation of the peninsula, much the same situation that existed from 1925 through 1927 when "...you could walk from the extreme tip of Long Island clear to Bad River" (Burnham, 1930). We learned from local residents that Long Island had not been a true island for a number of years prior to 1987. This status certainly differed from the situations such as in 1840, 1870, and 1891 when severe storms created gaps and separated the outer four miles of the peninsula as "Long Island." Burnham (1930) recounts engineering efforts to fill the storm-created gaps with "stone, slabs, and other material," but the materials "quickly disappeared into the sand." Long Island has apparently been changed from being a true island to a continuous part of the peninsula many times and usually persists under one status or the other for a few decades. When Long Island is connected to the mainland, the term "barrier spit" is appropriate; when it is separated from the peninsula, the term "barrier island" is appropriate. Both features of this size are rarely found in fresh water. Like marine barrier forms, the features are very dynamic. The destructive processes of storm waves serve to breach the peninsula from time to time, and the constructive process of sand deposition by longshore currents helps to slowly fill the gaps.

One can access Long Island from boat launches at or north of Washburn, from Bayfield, or from the marina at La Pointe on Madeline Island. This guide assumes an entrance at Washburn and a crossing directly east to the peninsula. Chequamegon Bay, between Long Island and Oak Point, is mostly less than 30 feet deep. The bay is an area frequented by sport fishermen.

Waters become very shallow at considerable distances from the spit, particularly in the area of Long Island. On the southwest side of the peninsula are several shallow estuaries that have an abundance of aquatic plants and water birds. Elsewhere along this shoreline are pleasant sand beaches. The water is shallow and the expanses of sand reveal shoaling, longitudinal, and bifurcating ripple marks much like those seen in the Precambrian rocks on Devils Island. Some areas of the peninsula are densely forested with evergreens, willow and birch, and eagles sometimes perch in plain view on the higher trees. The roots of the trees are exposed in places along the shoreline in extensive tight entanglements somewhat reminiscent of mangrove roots in tropical marine environments. The density of the entanglements here suggests that these roots are also agents that help to trap and bind the sediment. When storms wash away soil from roots, the trees here are much less stable than their marine relatives, which have developed specialized root systems custom fit to survival in a shoreline environment. The trees of Long Island are terrestrial species that have evolved to survive in temperate climates on dry land. They have never specialized to develop the root systems required to better survive conditions on the shores of a barrier island or sandspit. Thus, fallen trees cover many areas of the shoreline. On some parts of the shore, the beach is covered with wood chips and bark bits as the debris from trees joins with the sand transported by longshore currents.

The broader western part of Long Island is composed of several sand ridges. Apparently, the older part of the island is nearer the bay side (Collie, 1901). Small sand dunes and wind deflation blowouts are present. In places, grasses stabilize the dunes. As you approach the western tip of Long Island, the Chequamegon Point Lighthouse becomes visible. This is one of two lighthouses now on Long Island, the other being the La Pointe Lighthouse about one mile to the east on the northern shore. Arthur MacLeod, a prominent contributor to the Washburn Historical Society, recounted to us that the two lighthouses were once joined by a sidewalk only a few inches wide and that the lighthouse keepers traveled by motorcycle between the two stations. Dense overgrowth has obscured evidence of the sidewalk, and prolific stands of poison ivy around both lighthouses discourage investigation in summer. The several-mile uninterrupted sand beach along the northern edge of Long Island is the most extensive beach in the Apostle Islands and is an obvious attraction for hiking, swimming, and picnicking. Low wave-cut bluffs between the two lighthouses reveal extensive lag deposits. Farther east, along the northern coast, the height of the island diminishes into the area labeled "uncovers" on charts and topographic maps. This area reveals many low dunes and broad dry channels, which are active when washovers during storms carve away the sand and breach the spit every few decades.

The word "Chequamegon," which applies to the local bay, to the local lighthouse, to the occasional point on the peninsula nearest Long Island (when it intermittently becomes a true island), and to the most abundant geological formation in this national park, has been ascribed to several origins. According to one legend, the term comes from Shagewamekong, "soft beaver dam." The Indian hero Winnabozho constructed the peninsula to form a bay in which to trap a great beaver. During the frantic construction of this trap, great clods of earth were scattered and a number fell into the lake to form the Apostle Islands—a picturesque alternative explanation to the geological history we have presented here! Burnham (1930) notes that the general origin of the term possibly lies in the ancestral words from which "Chequamegon" was derived, words that apparently refer to a low area covered with water. Though the term was initially intended to refer only to this tiny peninsula, it eventually expanded in meaning to include a vaguely defined district encompassing all of northern Wisconsin. Chequamegon Point became the central settlement for traders, hunters, and war parties. "All roads began at Chequamegon Point" (Burnham, 1930).

With respect to the barrier spit, Burnham (1930) recounts the documented Battle of Chequamegon Point, in which a band of 150 Sioux captured two Ojibwa youths. Upon perceiving a rescue party, the marauders killed and scalped the youths. However, the rescue party they detected was only one of two deployed by the Ojibwas, and the Sioux, fleeing toward the mainland, quickly found themselves penned in from both sides on the narrow peninsula. Only two Sioux escaped alive. An account written in 1850 noted that the scattered bones of the Sioux were still visible on the sands of Chequamegon Point.

Madeline Island (Not Part of Apostle Islands NL)

Map References—Madeline Island; Madeline Island, Amnicon Point, and Bayfield 7 1/2' Quadrangles

Madeline Island, an island with a permanent population, lies outside the boundaries of the Apostle Islands National Lakeshore. However, its size and proximity merit it a few comments. This island is the largest of the Apostle Islands and can be accessed easily by automobile or boat. While making the ferry crossing from Bayfield to Madeline Island, you can get a good view of the highlands of the Bayfield Peninsula, which are Chequamegon Sandstone covered by glacial till and glacial outwash. The realization that the glacial ice overrode the peninsula above (see Fig. 6) gives perspective on the amounts of ice that once were present here.

Madeline Island, like most of the Apostle Islands, is underlain by Chequamegon Sandstone. Outcrops are not common because of the thick deposits of glacial till and outwash that cover the island. A few channel sandstones are exposed at the west end of the northeast-facing shore at the northern tip of the island, south of Devils Cauldron marked on the Amnicon Point 7 1/2' Quadrangle. More of this bedrock is seen along Big Bay Point and farther south along the southeastern-facing shore near the southerly end. Beautiful sand beaches occur along Amnicon Bay, along Big Bay, and at the sandspit, Grant Point, on the southeastern end of the island. Collie (1901) noted that the spit at Grant Point was once over 5,000 feet long, but had been greatly reduced in size by the time of his study in 1901. Big Bay State Park and the town of La Pointe's Big Bay Island Park preserve a baymouth bar built from sand deposited by longshore currents. Behind the barrier island is a shallow lagoon and an extensive wetlands. The Madeline Island 7 1/2' topographic map reveals a second, older barrier bar (now simply an inland ridge of sand) east of the present lagoon.

Big Bay Island Park has a number of campsites on the highlands above the lagoon, and the trail between the bay and the campsites provides outstanding views of the wetlands. At the southeast side of Amnicon Bay another wetlands and lagoon (Bog Lake) was formed when a baymouth was built across a bay. Along the northern coast of Madeline Island, most of the shorelines are steep actively eroding bluffs of glacial drift. In a few sheltered bays are narrow sandy beaches. Landings on most of Madeline Island would be on private property (with the exception of the state and town parks and the marina at the south end of the island) and should be avoided except in emergencies. At a few areas within the northern end of Madeline Island, pits have been opened to extract sand for construction purposes. The sand is glacial outwash.

Manitou Island

Map Reference—Oak Island 7 1/2' Quadrangle

The island, elongate northeast to southwest, is almost 740 feet in elevation. At one time, the southwest edge of the island extended to the warning light, but waves eroded material until only a small island (Little Manitou) remained where the light is today. Strzok (1981) notes that Little Manitou once served as a site for nesting birds, but the remaining unconsolidated material was washed away by the Coast Guard to "establish a more permanent base for the warning light." Little Manitou probably looked a good deal like present-day Gull Island. The restored fish camp on the southwest side of Manitou Island is one of the major attractions to visitors here. However, the northeast side has the most interesting geological exposures.

The bedrock that underlies Manitou Island and forms the shoals in the area of former Little Manitou Island is entirely Chequamegon Sandstone. Exposures of rock are poor over most of the island, with the best outcrops occurring along the north end. As you approach the northeast side, some low cliffs of channel sands with cross-beds dipping northeast begin to become visible where they protrude from the cover of glacial till. Here and there, a boulder of granite or basalt, weathered from the glacial till, rests in the shallows or along the water line. A few boulders in the water reveal the shallow, flat pits characteristic of weathered-out clay galls. Spectacular outcrops are found near the northern end, where waves have exquisitely exposed two channel bars in three dimensions. The sandstone beds that appear steeply inclined (Fig. 30) are actually cross-beds. Some beds contain gravels.

In addition to this cross sectional view in figure 30, one may walk along the top of the bar (Fig. 31) and observe the cross beds decrease in dip toward the crest of the bar.

Remember that the dominant flow direction of braided streams that deposited the Chequamegon Sandstone was to the northeast. As you orient yourself and imagine the waters of the rivers moving in that direction, you can begin to see how the dip direction of the cross-beds indicates the flow direction of the streams.

Figure 30. Cross-beds in Chequamegon Sandstone on northeastern Manitou Island. These built up one over the other when a sandbar grew in the channel of the Precambrian river. The dipping angle of the cross-beds is about the same angle of natural repose in which the sands were deposited over the edges of the sandbar. Change in cross-bed dips reveals where a second channel was later stacked along the edge of the earlier channel. Today you may walk over the surface and on each side of the sandbar, which was deposited about one billion years ago. Such excellent three-dimensional exposures are not common.

Figure 31. Channel sandbar in Chequamegon Sandstone on northern Manitou Island. At this site, two sides plus the top of the sandbar are exposed along a boulder-strewn bench. This figure shows mainly the downstream edge of the bar where layer upon layer of steeply inclined cross-beds were built when sand spilled over the front of the bar (left to right) in the Precambrian river.

Michigan Island

Map Reference—Michigan Island 7 1/2' Quadrangle

Michigan Island is a long narrow island on which no bedrock is exposed. Spectacular bluffs of glacial till completely cover the Chequamegon bedrock and rise abruptly from the water line. The red color of the till arises from small amounts of iron oxide and is particularly striking when the island is viewed some distance from the west at sunset. The bluffs appear especially steep and high when viewed while crossing from Madeline Island toward the lighthouse on Michigan.

The northeasterly part of the island is of generally low relief and is poorly drained. Although no bedrock appears on Michigan Island, it is undoubtedly close to the surface, as is indicated by huge boulders that lie beneath the water along the entire eastern side of the island. Boulders make the trip up the eastern edge of the island a chore that requires constant attention; weather permitting, the west side is much safer.

There are no huge boulders weathering from the glacial bluffs, so the boulders in the shallows along the east side are likely bedrock. It is likely that the thick tills seen in the bluffs once covered the boulder-strewn platform on the east side; the high bluffs mark the progress of erosion up to the present. Erosion may proceed quickly on Michigan because the bedrock appears to be just below the present lake level, and therefore waves on all sides of Michigan Island encounter soft glacial till rather than resistant rock. The broad shoals shown on navigation charts around the north, west and east sides of Michigan Island reveal the amount of till stripped by erosion since the glaciers receded. The current area of the Island is probably half its original area. To the north, Michigan tapers to a rock reef

Figure 32. Sandspit on southern tip of Michigan Island. The sandspit is noted for its extent and its beautiful display of bleached driftwood.

Michigan

Parts of trees have washed up on the beach
at the southern end of Michigan,
a community of limbs
washed gray by Lake Superior,
silver in the sun.
Some stroke the sand in motion with the wind;
others pierce the sand or intersect at angles
or jut upward like strange obelisks,
a patternless pattern of gleaming silver
against blue.
Stonehenge has not such happy disarray
nor could one dare to play
upon those solemn stones.
But here we climb upon the sticks
to perch like herring gulls,
entwine our limbs with these limbs,
decorate with ourselves the silver spires
which seem to intimate obeisance to some god
who laughs,
who uses bones of trees to make a playground,
for whom sacredness is common delight
and not solemnity.
Ah, Michigan, lend us the strange happy god
who looks down on your southern beach.

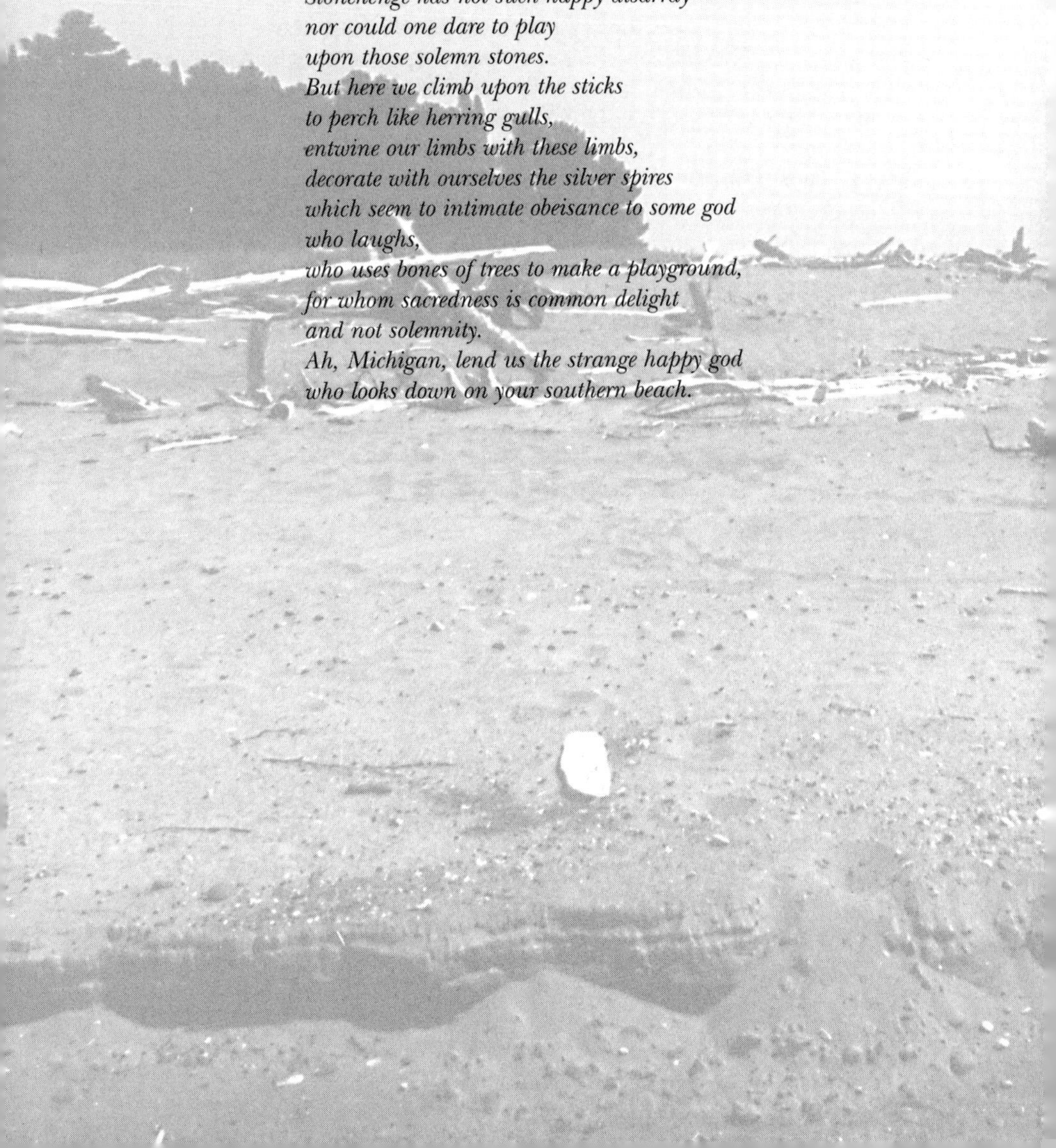

that runs from the northern tip past Gull Island. The reef is Chequamegon Sandstone bedrock with a thin cover of cobbles and boulders.

The large sandspit (Fig. 62) on Michigan Island is very attractive and worth spending time on (Fig. 32). Engstrom (1985) noted that the spit was growing on the south end but gradually getting smaller on the north end. Sand is being contributed to the spit from both sides of the narrow island, where steep bluffs of glacial till are rapidly weathered.

Examination of this till shows that it is extremely sandy and that it is exposed all the way to the water line. Wave action is the most important agent for eroding the bluffs and for moving the sand southward to the spit by longshore currents. Away from the sand beach, a zone of picturesque driftwood occurs along the dunes and foredunes. The foredunes have an abundance of dune grass and sandbar willow with some red maple and sage. A beautiful wetlands exists in the interior (see Fig. 62), a short distance north of the sandspit. One can access a view of it from the southwest shore of the island by simply hiking up over the low vegetated beach ridge above the narrow sand beach. Another wetlands exists at the northeast end of Michigan Island, where ponds feed a small perennial stream which exits near the tip of the island onto the rock reef separating Michigan and Gull Islands.

North Twin Island

Map Reference—Cat Island 7 1/2' Quadrangle

North Twin Island appears different from all other islands. An atmosphere of extreme age and durability seems to engulf the visitor, and its solitude adds to this pristine beauty. The entire island is surrounded by low, clean rocky cliffs (Fig. 33) that reveal no recent landforms such as beaches or slide scars. Because of its small size, remote location, and lack

Figure 33. Shore of North Twin Island. The shoreline lacks thick glacial deposits and thus is devoid of beaches, spits, or other evidence of recent processes.

North Twin

North Twin rests apart
beyond time
beyond the Islands
her brown sandstone hoary.
Having endured the waters
once
and again
and then again
she stands gracious and impervious
a sage who has long since
abandoned struggle for acceptance
an ancient one who has exchanged
passion for peace.
As we pass by in the bass boat
we greet her
as an old Indian grandmother
warm from many suns
heavy from holding much life,
at once solitary and selfless.

of a good anchorage, North Twin was probably not logged, as were the other islands. Thus, its stand of vegetation is old and profuse. Although access is not restricted, camping is not permitted here.

The shoreline is generally consistent in appearance all the way around the island. It is mostly sheer low outcrops of Chequamegon Sandstone. Although Strzok (1981) noted that the island "...is very similar in shape and foliage to Devils Island," these islands are geologically dissimilar. The shorelines are completely different. There are no sea caves on North Twin Island, only a single shallow reentrant along the east side. Views of cross-beds are abundant on North Twin. The tops of the low cliffs are wave-cut platforms covered with broken slabs of rock a few feet shoreward of the cliff edge. Glacial grooves may be seen here and there between the slabs. In places, stress from glacial ice has brecciated the upper portions of the rock. The cliff at the northern end is capped with a thin conglomerate.

Near the southwest side, glacial till lies on the rock, but not enough material is present to furnish sufficient sand needed for any spits or beaches. Instead, brownstone gravels litter most of the narrow southerly shoreline, although a small bench of laminated sand exists at the south-western end.

Oak Island

Map Reference—Oak Island and York Island 7 1/2' Quadrangles

Oak Island provides many attractions, and we highly recommended a visit. Features include rock windows, waterfalls, small sea caves veneered with moss, clay cliffs, sandspits, hiking trails, overlooks, beach ridges, and good campsites. Detailed soil maps were made by Cary, McDowell, and Graumlich (1979), but little bedrock geology was described by these workers. The island is large (over 5,000 acres) and is the tallest of the islands, with a maximum elevation of 1,081 feet At the high point (and at other areas also), you can reach beneath the thin leafy humus cover and touch clean fine sand, an indication that this was once a beach which was slightly emergent or just beneath shallow water. Recall from the section on general geology that Clayton's (1985) estimate of the high stand of lake water was 1,075 feet—very close to the elevation of the high point here at Oak Island. Falling lake levels left several beach ridges down to an elevation of about 1,000 feet. Below 1,000 feet, some rock-strewn terraces between an elevation of about 900 and 800 feet may represent remnants of cobble beaches. Trails leading to the high point and to other sites on Oak Island are particularly pleasant. The trails are accessed from the National Park Service dock on the western side of the island. The island is beautifully forested and well drained. Although the tall trees prevent a view from the highest point, a trail leads to a spectacular overlook at the north end, from which many islands can be seen.

The shoreline around Oak Island is varied and interesting. If you leave the west National Park Service dock and turn northward, you soon begin to encounter a few channels in the Chequamegon Sandstone along

Figure 34. Clay banks on the northwestern tip of Oak Island. The clays are laminated. The laminae were possibly laid down in a small kettle lake within glacial outwash. (Kettle lakes form when immense blocks of ice are buried, melt, and form a collapse depression that fills with water to form a lake or pond.)

the point where the shoreline trend changes from northwest to northeast. As you round the point to the northwest beach, gravels composed of native Chequamegon Sandstone as well as cobbles of basalts, gneisses, granites, and quartz eroded from glacial drift are present. A bit farther northeast, you encounter the "clay banks" (labeled on Oak Island 7 1/2' Quadrangle). Portions of these clays are thinly laminated with alternating light and dark layers that may be seasonal varves deposited in ice-age ponds or small lakes associated with glaciers (Fig. 34). The plastic nature of these sediments is evident from touching them. In places, small liquid mudflows may be seen spilling out over the beach sand (Fig. 35).

Oak Island may be recognized easily at a distance by a prominent reddish bluff (Fig. 36), on which are exposed several types of glacial deposits. As you approach the promontory along the northern end of the island, you see definite evidence of glacial outwash deposits. Outwash differs from till in that outwash is deposited by flowing meltwaters. Water-laid deposits, which tend to be sorted and stratified, reveal current forms such as cross-beds (Fig. 37).

After you pass the promontory, the route enters the North Bay. The bay is sandy, and after rains, a swath of water clouded with silt and clay can be seen in the bay leading back toward the promontory. An attractive campsite lies near a stream at the eastern end of the bay. A prominent rock window, virtually a natural bridge, called "Hole-in-the-Wall" (Fig. 38)

Figure 35. Miniature mud flows at the "clay banks" of Oak Island. These are formed where ground water saturates the clay until it liquefies and flows.

Figure 36. Prominent bluff in glacial drift on Oak Island. The bluff, being scaled by adventuresome climbers (see arrow), is best approached by a trail, which leads from the park dock to an overlook at the top.

Figure 37. Stratified outwash deposits on northern Oak Island. The stratification is caused when running water (in this case, meltwater from glacial ice) transports and deposits sand and gravel.

marks the northeast corner of the island. A close examination of the bridge from below will show that it is an isolated rock precariously key seated between two joints. The northwest side of "Hole-in-the-Wall" reveals beautiful foreset cross-beds, which were buried by a complex array of channels with scour-and-fill cross-beds. A short climb onto the promontory gives you a beautiful view of the lake behind a craggy twisted pine that appears like a bonsai (Fig. 38 inset).

If you proceed south, slowly and close to shore, from "Hole-in-the-Wall" in a small dinghy or kayak, you'll encounter many subtle but beautiful features along this rocky shore. You can see conglomerate beds, often overlain by sand channels with a complex interlacing of scour-and-fill cross-beds. A few slabs on shore reveal undulatory ripple marks. Farther south, a narrow strip of sandy beach lies at the base of rock overlain by glacial deposits. These continue for about a mile. Along the southern third of the island, facing east toward Stockton, rock again appears. Here, streams flow from the island; at several places, small waterfalls can sometimes be seen. In places, the moisture maintains a brilliant green cover of moss on rock, which sometimes resembles miniature well-groomed lawns. The careful observer will see some straight-crested ripples on a few rock surfaces here, evidence of very quiet water—possibly ponded water between sandbars in the old Precambrian streams. One particularly attractive waterfall tumbles over a reentrant into the lake. Paddling a small boat under this waterfall, you can see through the clear water many sandstone cobbles with beautiful liesegang color bands that range from many shades of brown to tan and white. These color bands are caused by selective

Figure 38. "Hole-in-the-Wall" on Oak Island. This feature is the product of undercutting of a channel sand along two nearly vertical joints. Inset is view looking north from cliffs of Hole-in-the-Wall. **Avoid stepping onto the natural bridge that forms the rock window.** *It is unsafe to walk on and could collapse.*

deposits of iron in the rock, probably iron diffusing through ground water. Seen lying together in a sunlit pool beneath the waterfall, these stones are exquisitely beautiful. Once out of the water and dry, however, they appear dull and less striking.

The southeast corner of Oak Island is unusual because the sandstone cross-beds dip southeast, thus indicating that they were deposited by currents that flowed at about right angles to the normal northeasterly trend of the Precambrian streams. This almost certainly represents deposition along a bend in the ancient river, which probably flowed northeasterly out of the area of the Bayfield Peninsula, swung southeast across Oak Island and then swung back northeast toward Stockton Island. This perhaps accounts for the orientation of the longest axis of Oak Island at approximately right angles to the dominant NE-SW elongation typical of most islands in this archipelago.

After rounding the southeast corner of Oak Island, you proceed along a south-facing shore that has some slumps, and is mostly covered by glacial drift. This drift and that lying along the southwest-facing shore supply sand to the attractive sandspit at the southwest corner of the island. Several good campsites are present as well as vault toilets. In accord with its position close to the southwest shore of Lake Superior, the sandspit is slowly

Oak

At 3:00 it darkened to half-night
and we knew it would rain.
Gliding along the trail like ghosts
we caught the rain–
when it came–
second-hand, as it dripped
from heavy leaves;
our rain ponchos glimmered strangely
in the wet half-glow
and we moved through the woods
of Oak Island
as three faint lights–
three lights in a dark and muddy wood
moving forward in the rain–
a cocoon of light in a dark wood,
the white birch and one great oak
rising up and touching.
In the night,   late,
the rain stopped and
fog snuggled down into the cleavages
of the hills, sending a milk-breath
sigh of content
low onto the edges of the lake.
At campsite 11, along a narrow strip of beach,
we woke knowing it was good
to be there in that place.

shrinking, likely as a result of the locally deepening lake water engulfing the low-lying sandspits and beaches along southernmost Lake Superior (Engstrom, 1985). Post-glacial rebound is one factor that contributes to this deepening. Hartford and Tanner (1976) note that changes in amounts of precipitation and water diversion from the Hudson Bay may contribute to rising lake levels. Geomorphic evidence indicates that the encroachment of lake waters onto these shores has been occurring for some time (see Collie, 1901), certainly prior to any water diversion projects.

Otter Island

Map Reference—Oak Island and Rocky Island 7 1/2' Quadrangles

Otter Island is completely underlain by the Chequamegon Sandstone. Although outcrops may be found in places around the northern two-thirds of the island, good cliffs are seen only along the north and north-west edges. Along the western edge, there are a number of recent rock-falls, as revealed by a litter of boulders and fresh scars marking the place from where they fell. At the north end, the lake waters have eroded a wave-cut platform with an accompanying beach of cobbles, boulders and large driftwood logs (Fig 39).

Many joints are seen as long straight cracks going through the rock and as flat smooth faces along the exposed cliffs. Although no streams appear on the topographic maps, a strikingly beautiful waterfall, one of the nicest we saw in the park, was present in mid-August of 1986 (Fig. 40) on the north shore of Otter Island. It was gone in 1996, and possibly was just an ephemeral feature. The north shore appeared greatly changed as

Figure 39. Cobbles and boulders on wave-cut platform, Otter Island, show the force of water present at times unlike when this photo was taken. Small scour "potholes" are present, and the large boulders are rearranged each year as result of strong winter waves.

Figure 40. Ephemeral waterfall on northern Otter Island.

result of rockfalls when we visited there again in 1996. Beaver inhabit the islands, and the occasional failure of a beaver dam may account for ephemeral waterfalls seen during dry periods.

The rocky shorelines of Otter had so many rock falls between 1986 and 1996 that in places it was impossible to match any photos taken in 1986 with views displayed a decade later.

Outer Island

Map Reference—Outer Island 7 1/2' Quadrangle

The eastern and western sides of the island are three different geological worlds. The oldest is the eastern side which is almost entirely one-billion-year-old Precambrian Chequamegon Sandstone bedrock. The exposures are best summarized as wave-cut platforms, which appear as flat-surfaced shelves and benches. Although there are no sea caves, the shoreline is still impressive. Excellent exposures of stacked channels exist (Fig. 41), as well as massive foreset beds of giant sand ripples (Fig. 42). The east side of the island still clearly displays the conspicuous structure in the sandstone (Fig. 43) depicted in the frontispiece of Thwaites (1912) as an "up curve." It probably results from settling that occurred when the sands that formed into Precambrian sandstones were still soft. Such convoluted beds (Fig. 44), which are sedimentary beds contorted by subsequent settling while soft, are exhibited in some vertical exposures, especially where the exposures coincide with joint faces. When weather conditions are good, opportunity exists to access the cliffs and hike over the flat wave-cut benches. One ledge reveals a cobble conglomerate, perhaps the coarsest conglomerate in the Apostle Islands (Fig. 45). The size of the cobbles reveals the power of the currents in the northeastward-flowing river.

Figure 41. Channel sands viewed on the east side of Outer Island. The flow direction is to the right. The stacking of thin sands, each overlying sand representing subsequent deposition in a channel, is exemplary of sand accumulation in a braided stream environment.

Figure 42. Thick planar foreset beds on eastern shore of Outer Island. These thick foreset beds represent rapid deposition in braided streams. Together, these inclined foreset beds have built the large sandbar seen here. The bar likely accumulated quickly, perhaps during a flood.

Figure 43. The "up curve" structure photographed by Thwaites, along east side of Outer Island in a field season prior to 1912, is still clearly visible along a vertical joint face nearly a century later.

Figure 44. Convoluted beds on eastern shore of Outer Island. Convoluted beds represent settling and slumping of the sands while they are still soft and saturated with water. They occur in many modern environments, including braided stream environments. Convoluted beds are seen in a number of outcrops of the Bayfield Group.

Figure 45. The large cobbles in this conglomerate on this flat wave-cut bench of Outer Island resemble a cobblestone walkway set by a master stonemason. To move such cobbles takes a lot of force generated by flowing water. This deposit represents an abnormally strong flow event in the Precambrian streams that placed the cobbles here.

Figure 46. Beach along the west side of Outer Island sandspit. The extensive beaches are products of strong waves and a local abundance of sandy glacial till.

Figure 47. Stratified sand and gravel deposits on southeast side of Outer Island, exposed in bluff about 20 feet above level of present lake. Cross-beds at this exposure dip in several directions and indicate that these materials were probably part of an old sandspit that formed here when the lake level was higher.

Undoubtedly, sands of the Bayfield Group were deposited well beyond the present northeasterly limits of these islands.

The western exposure, where reddish drift deposits are exposed in bluffs nearly one hundred feet high, represents the glacial age; the ends represent the modern era of geological processes, which are continuing to build a beach at the northern end and the largest sandspit in the islands at the southern end. The sandspit itself is worth the trip.

The glacial bluffs are made up of sandy drift, apparently similar to that found on Michigan Island. Little detailed work has been done to date on these glacial deposits. The topography shown on the quadrangle map is streamlined in a trend about N45°E in accord with the direction of ice movement of the most recent stage of glaciation. Clayton (1985) labels these on his regional map as several drumlins (elongate hills of unconsolidated material streamlined in a particular direction from being overridden by moving glacial ice). Most of the material is probably a till, although it is presently uncorrelated with till materials on the mainland.

The beaches of the sandspit on the south end (Fig. 62) are exquisitely beautiful (Fig. 46). Data from Engstrom (1985) show this sandspit to be lengthening, but generally building on the east side and receding on the west. A campsite and a vault toilet are present on the sandspit among trees and blueberry bushes near the east side.

A thin sand beach in the cove on the northern end just east of the lighthouse is apparently a good illustration of rapid processes. In 1985, Taber Allison, then a researcher from the University of Minnesota, described how

Figure 48. Armored mudballs from southern Outer Island.

he witnessed the beach scoured away in a few hours' time by a single storm, and saw it built back again within a few days. Some sand is also present in Austad Bay, depending on recent weather conditions. In contrast, the beaches of the sandspit at the southern tip of the island are less transient.

When lumbering was an active industry here, a railroad operated on the island and two spurs were present on the sandspit. Hemlock was loaded off the west dock and hardwoods off the east side. You may see remnants of a large boat lying high on the west side of the sandspit. This is probably the sunken boat "Faithful," owned by the former Lullaby Log Company (Middleton, 1983).

If you look at the bluffs on the southeast side immediately north of the sandspit, you will notice two layers of gravel interbedded with stratified and cross-bedded sand (Fig. 47) overlying the clay beds. At the base of the bluff, a peculiar sedimentary feature, armored mud balls (Fig. 48), can be found. These form when a layer of clay is eroded by waves and broken into lumps. The sticky coherent clay is rolled about on the beach and picks up pebbles that eventually form "armor" around the mud lump, which helps to keep it from being dispersed.

Extensive wetlands and ponds are present on the northwest side of the sandspit. Much of the interior of Outer Island is marshy and poorly drained. These conditions result from the nature of the till deposits that cover the island and possibly from beavers constructing dams that cause further impoundment of water.

Care is needed when visiting Outer Island. The length of the island almost dictates that whatever weather you begin with will change before you can reach the other end. The exposure dictates that these changes will be experienced to their full severity. The eastern side offers almost no shelter beyond a few very narrow coves. The only good anchorages are at the north or south ends.

Outer

During summer we waited to get to Outer.
Wind roughed up the water for days
and we remembered the Pretoria, the steamer James H.Hoyt,
the George H. Van Vleck, the Ontario.
Twenty-five miles from Bayfield, Outer lingers
there among her boulders
as ghostly as the vessels she wracked.
Four miles east of Cat, three miles northeast of Stockton,
seven north of Michigan and Gull–even these distances are
unsettling,   for a bass boat.
Fear came into the brain like the sound of breaking glass.

Spring in Platteville I wait still to get to Outer,
not the 8000 acres of sand dunes and clay
twenty-five miles from Bayfield,
but the outermost island which like Outer lies somewhere
in desperately shallow waters fraught with boulders,
that ghost with a half-familiar smile
and white fingers that beckon,
the somewhere both necessary and life-threatening.
I wait to get to the outermost place,
knowing I must touch this ghost once removed from touching,
even while I hear in my brain the tinkle of breaking glass.

Mind is not the road to the outermost island;
it circles back upon itself, a golden ring.
Nor will body yield the destination;
it circles back upon itself, a golden ring,
though the feet which tread face always east.
There is no road to the outermost island
though some have found the place and named it
"gachiishquaguindag miniss."

Raspberry Island

Map Reference—York Island 7 1/2' Quadrangle

Raspberry Island is about 300 acres in size. Cary, McDowell, and Graumlich (1979) produced a soil map of this particular island. The lighthouse together with its famous flower and vegetable gardens is located in the southwest corner at the top of a steep bluff of glacial drift. For a time, gabions (wire baskets filled with rocks) filled with glacial erratics were once placed in critical areas to try to slow down wave erosion. Neil Howk, assistant chief of interpretation and education at Apostle Islands National Lakeshore explained that in 2002 and 2003, a major slope stabilization project replaced the gabions. Workers recontoured the slope and built a rock revetment made from broken local stone and erratics along the shoreline in front of the light station. The revetment extends approximately 210 feet northwest and 130 feet east of the lighthouse docks. It currently stabilizes the slope and protects the historic buildings. The top of the armoring riprap stone is five feet above the dock surface (Fig. 49). Above the riprap, earth fill was added to build the slope in front of the lighthouse, and the final slope was seeded and replanted with native plants.

As you proceed from the lighthouse along the northern shore, you eventually encounter some low outcrops and boulders of the Chequamegon Sandstone—exposures of channel sands viewed parallel to the direction of flow. They exhibit northeast-dipping cross-beds and megaripple forms; a few subarkosic pebbly layers are present with several thin foreset beds. The uneven channel bedding becomes covered with glacial till in the northwest-facing bay. The bluffs of glacial till have been

Figure 49. Slope stabilization with riprap at the base of Raspberry Island Lighthouse (photo from National Park Service, Apostle Islands National Lakeshore).

Raspberry

I wanted a green bean
from the garden at Raspberry
so Bob the Lighthouse Keeper
lifted me up and extended me out
over the sea of green like an oar
and I delved in the depths and
came up with a bean.
He wouldn't let the toe of my tennis shoe
disturb the row, for
of the gardens of the lighthouse keepers,
Raspberry's has been.

Once during a Celebration of the Land
Ben Logan said, "I don't think,
when it comes to the land,
you can talk about beauty and ugliness.
You can't make that distinction."
A fellow writer said,
"Hell, Ben,
you know ugliness on the land
when you see it
as well as I do."
Then a white-haired woman asked
what is the land and we couldn't say;
we had to call it X.

In the garden
the single weed grows with grace.

In our shallow wisdom
we have presupposed good and bad,
beauty and ugliness.

Zen poet Shinkichi Takahashi

Figure 50. Sand beach along the cuspate foreland on southeast side of Raspberry Island. Bay to northeast is an anchorage favored by sailboaters (photo from National Park Service, Apostle Islands National Lakeshore).

incised by gullies, and small slump blocks with living trees have recently slid downslope onto the shore. The shore itself has a narrow beach of sand and coarse gravel. At the northeast end of the bay, some stratified glacial outwash sands are visible. Farther ahead are boulders of Chequamegon Sandstone covered with mosses of many shades of green. Low ledges exhibit cross-beds. As you pass the point on the northeast side, you can see the tall bluffs of Oak Island in the distance.

Proceeding southwest along the southern coast into the bay, you will notice a decrease in size of the particles along the beach—from boulders, to cobbles, to pebbles and sand. This decrease reflects the winnowing of fine sand from the glacial bluffs along the northeast end and the movement of the sand down-current into the sheltered bay. The bay, with its deep sandy bottom, is a favorite anchorage for sailboats. At the southwest end of the bay is the Raspberry Island sandspit (Fig. 50), a cuspate foreland with a nice sand beach and a marshy interior. Two researchers, Engstrom (1985) and Middleton (1983) studied this sandspit. Engstrom's study indicated that the spit was gradually decreasing in size, but Middleton noted that it had lengthened by about 100 feet between 1938 and 1978.

Rocky Island

Map Reference—Rocky Island 7 1/2' Quadrangle

The bedrock underlying Rocky Island is Chequamegon Sandstone, but it is almost entirely covered by glacial till. Despite its name, this island has little exposed bedrock, and almost all of Rocky Island is covered with glacial drift. The soils map produced by Cary, McDowell, and Graumlich

(1979) and the regional map of Thwaites (1912) both show the only bedrock outcrops to be at the north shore near the center of the island, just west of the narrow neck. The island gets its name from the gravels (mostly cobbles and pebbles) that fringe the southernmost and northerly portions of the island. At the extreme northeastern end of the island, a shoal of bedrock extends beneath the water for some distance. Joints may be seen through the clear water as well as a litter of very large sandstone boulders.

Most of the western shore of the island consists of sheer bluffs of glacial till. The bathymetry indicates that Rocky and South Twin Islands were probably once a single large island with a deep bay at the south end. Even today the water between the narrow part of Rocky Island and South Twin Island is in places less than 10 feet deep. Rocky Island has a beautiful sand beach along most of the eastern side and a large sandspit at the southeastern tip. Middleton (1983) compared aerial photographs taken in 1938 and 1978 and found no area change in the sandspit over the forty-year interval, but Engstrom (1985) reported a decrease in size of the spit over the decade preceding 1985. The broad shallow shelf that extends around the northern half of the island to include South Twin Island is probably responsible for sheltering the eastern side of Rocky Island and for permitting sand to build wide beaches.

Sand Island

Map Reference—Sand Island 7 1/2' Quadrangle

Sand Island is a large, low island (Fig. 51) that is poorly drained and heavily overgrown, making hiking difficult. However, the shoreline itself provides such a wealth of features that this island is one of the most attractive of all. It lies within an area of "sands"—Sand Point, Sand River, Sand Bay, Little Sand Bay and Sand Island. On the highland southeast of Little Sand Bay are sand and gravel pits. This "area of sands" owes its origin to glacial melt waters that actually transported large amounts of well-sorted sand and gravels into the area.

When weather conditions are good, the island is readily accessible from the Town of Russell public boat launch near the National Park Service dock at Little Sand Bay. However, a sandbar extends roughly north-south from the southeast end of Sand Island almost to the mainland. Water there is less than 6 feet deep—sometimes less than a yard deep. In moderate east or west winds, waves pile up at this barrier, posing a threat to small boats. The barrier is more than just an accumulation of sands. A bedrock ledge, about 30 to 35 feet below the lake surface extends from the mainland to Sand Island and beyond northeastward to the Sand Island Shoals. Local streams such as the Sand River, which have their watersheds in sandy glacial materials, carry large amounts of sand to the lake. This sand accumulates on the rock platform.

Sand Island is interesting geologically because both the Orienta and the Devils Island formations are present. Thus, many features can be examined here that you would normally be able to see only by traveling

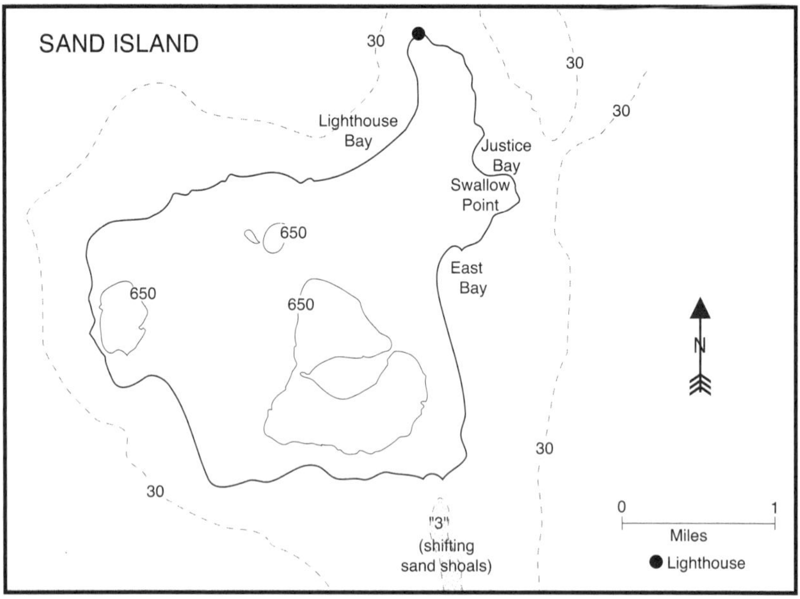

Figure 51. Map of Sand Island showing general elevations (feet) and bathymetry (dashed lines). The shallow sand bar on the south side of the island is not only a barrier to larger boats but serves as a place where waves build up that can sometimes pose problems for small open boats.

farther offshore. If you leave Little Sand Bay and proceed to the southeast corner of the island, you will encounter a few outcrops of the Devils Island Sandstone. Most of the outcrops are low cliffs. As you proceed north along the shoreline and enter East Bay, the rock outcrops become covered by sand. East Bay offers good anchorage and shelter when winds are from the west. A little farther north, the Devils Island Sandstone again outcrops near Swallow Point. The Devils Island Formation is the one that is associated with the famous sea caves on Devils Island, and here at Swallow Point are a beautiful set of sea caves (Fig. 52). These are best visited in early morning, when the sunlight enters the east-facing caverns. The caverns, lower than those on Devils Island, are still large enough to enter with a small boat. Intricate sculpture, windows (Fig. 53), and the colorful nature of the sandstones make this a particularly enjoyable stop. Wave-formed ripple-mark casts may be seen in the roofs of the caverns. Outside the caverns, the ripples themselves may be studied by climbing onto some of the flat ledges nearby.

After leaving Swallow Point, you leave the cliffs and pass northward into Justice Bay, termed by Strzok (1981) as "one of the most beautiful among the islands." The contact between the Devils Island Sandstone and the underlying Orienta Sandstone lies just north of the center of the bay, but is concealed beneath sand and glacial clays at the lake level. The rocks in the cliffs to the north are Orienta Sandstone. Continuing north, you

Figure 52. Caves in the Devils Island Formation at Swallow Point on Sand Island (see also discussions under Devils Island section).

Figure 53. Windows at Swallow Point on Sand Island in Devils Island Formation as viewed from a boat inside the cave.

can see the more massive nature of the sandstones of the Orienta Formation in contrast to the thin-bedded Devils Island Formation. The sands in the Orienta were deposited by the northeastward-flowing Precambrian (one-billion-year-old) streams. One of the best places to examine the Orienta is at the lighthouse at the northern tip of the island. The lighthouse was completed in 1881 from the native "brownstone," likely quarried from Basswood Island. It is the only lighthouse in the islands constructed of native stone (Fig. 54 inset), and this beautiful structure is

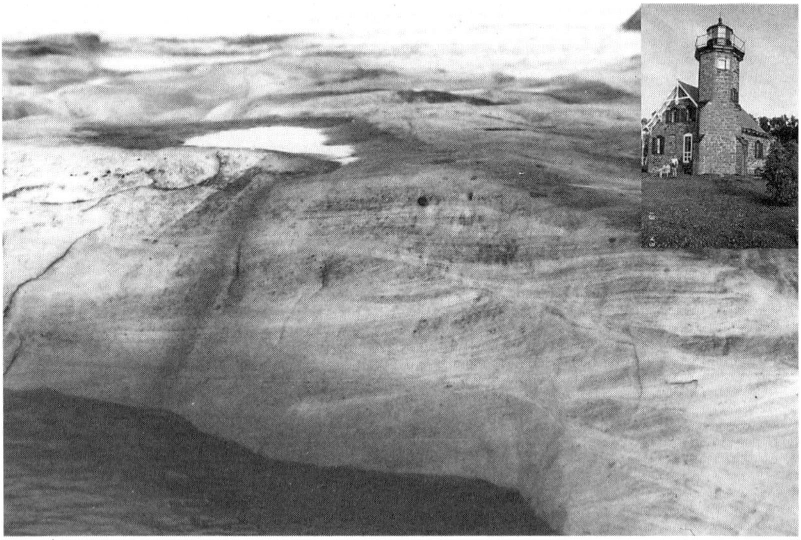

Figure 54. Scour-and-fill cross-beds in Orienta Sandstone outcrop just below Sand Island Lighthouse (inset). These were formed by northeastward-flowing streams about one billion years ago.

worth touring. Below the lighthouse, a large shelf of light-colored tightly cemented sandstone is exposed about a foot above water level. Overlying this sandstone, farther back toward the island, is a bed of red laminated shaly sandstone. Close examination reveals that the lighter-colored massive sandstone contains many scour-and-fill cross-beds (Fig. 54). Some parting lineation is also present. These current-formed structures are useful for deducing the northeastward direction of flow of streams that deposited the sands.

Another perspective is gained by looking at the beds in a larger context. If you travel along the island from some distance east of the lighthouse and then on past the lighthouse, you will see wide undulations along the contact of these two rocks. The undulations are probably giant ripples formed as the underlying sand was piled up in regularly spaced bars along shallow river channels. The laminated reddish sediments were then deposited onto this sand, and they filled in the irregular undulations. Recent erosion has stripped away the softer overlying layer to again reveal the undulating giant ripples.

Lighthouse Bay to the west of the lighthouse contains a beautiful sand beach. Summer sunsets viewed from here are exquisite. On the upper shore face, sands accumulate in alternating light and dark-colored layers (Fig. 55). The dark layers are concentrates of heavy minerals such as magnetite (Fig. 55 inset) that concentrate together during waning conditions of storms. The dark minerals are denser than the lighter quartz and feldspar, so as waves attenuate, the denser minerals are dropped selectively into a black layer, while the lighter colored ones remain suspended a bit longer.

Figure 55. Lag deposits on upper Lighthouse Bay beach on Sand Island. Dark layers are black magnetite, rutile and ilmenite. A magnet (inset) detects the many grains of magnetite from within the dark layers.

At the west end of this bay are some spectacular potholes (Fig. 56). They reside at just about normal lake level and sometimes will be well exposed and other times will be below water depending upon the annual elevation of Lake Superior, which commonly can vary by as much as three feet. These particular potholes, some large enough to sit in (Fig. 56 inset), require considerable erosion by a large stream or river with waterfalls and rapids. Their presence here poses a problem, because such a river requires a large watershed and a steep gradient, and as noted above, this island is low, provides no credible catch basin for such a large river and no such river is here. We believe that these particular potholes represent the time when the glacial Lake View Advance (Fig. 7C) receded. The large drainage basin for the stream that formed these potholes would have rested on glacial ice itself. The river flowed south down a steep gradient of ice provided by the glaciers' southern margin. Water gushing down this front carried hard cobbles and boulders and churned them around and around on the Orienta Sandstone, until they wore deep recesses—the potholes—into the sandstone. The ice, the river, and its course have been gone for thousands of years, but the potholes remain. Farther south in Wisconsin near Baraboo, the terminus of an earlier, more southerly glacial advance produced potholes in a similar fashion on the quartzite bluff above Devils Lake. These can be seen along the Potholes Trail in Devils Lake State Park.

As you proceed west from Lighthouse Bay, many exposures of Orienta Sandstone are seen. Channel cuts and fills are common, both here and farther along the west side, and cross bedding is common. The rock outcrops continue all the way into West Bay, which serves as a good shelter for

Sand

A devil has escaped Devils Island.
I saw it on Sand.
It danced out of a sunset of 100 pinks
and its shadow darkened the birches and pines opposite
making them fearsome.
It fled to the sea caves near Swallow Point
where I heard it glugging and gurgling
sometimes moaning.
I searched for it there in the caves
and I found it
and am haunted by it.
Whatever things I hold in my hand now
are tinged with 100 pinks
or shadowed.
What I choose to do now
vibrates with gurgling
and moaning.
Could I but loose myself
like light at dawn released into a larger world
could I but unhaunt myself
could my body reach outside its Sand
and learn to be without its hundred pinks
(is it audacity to think of it?
to yearn for more than this a place to be?)
Could I but go from Sand into a place
a larger world
a place of no place
a place beyond the place of no place
to know beyond the sound of no sound
the color of no color…
What devil is this
escaped from Devils Island
which makes me dream the freedom
of emptiness?

Figure 56. Potholes viewed near west end of Lighthouse Bay, Sand Island, in 1988. When lake levels are normal to slightly higher, all of these may be below water.

small boats when winds come from the east. There, however, boulders are scattered beneath the water, some close enough to the surface to contact the outboard prop of even a very small boat. The bluffs in West Bay are glacial drift. Both the sheer drop of the bluffs and the number of fallen trees testify to active erosion of the shoreline.

You can return to Little Sand Bay after proceeding along the south shore of Sand Island. Realize that the docks on the south side of Sand Island are still private and are not for public use. If there are several hours of daylight left and water conditions are good, a crossing directly south to Sand Point and thence westward to Mawikwe Bay Caves (see "Sheltered Water Tour—Bark Bay to Little Sand Point") would be an attractive option.

South Twin Island

Map Reference—Rocky Island 7 1/2' Quadrangle

Summer sunsets viewed over Rocky Island from South Twin Island rival those of anywhere in the world. In terms of appearance and geological history, South Twin Island is more closely akin to Rocky Island than any other island. In fact, bathymetry of waters here indicates that the two were joined at some time in the past. Although no exposures of bedrock appear on the island, bedrock may be seen at the north end below the shallow clear water. There, very large boulders and joints reveal that the Chequamegon Sandstone is covered by only a relatively thin layer of till. Glacial drift is exposed around the entire island. Along the beach at the north edge of the cuspate foreland, sands may be seen overlying coarse gravels that are either glacial outwash or old beach gravel. Some stratified sand outwash is also exposed in the low bluffs at the south end of South

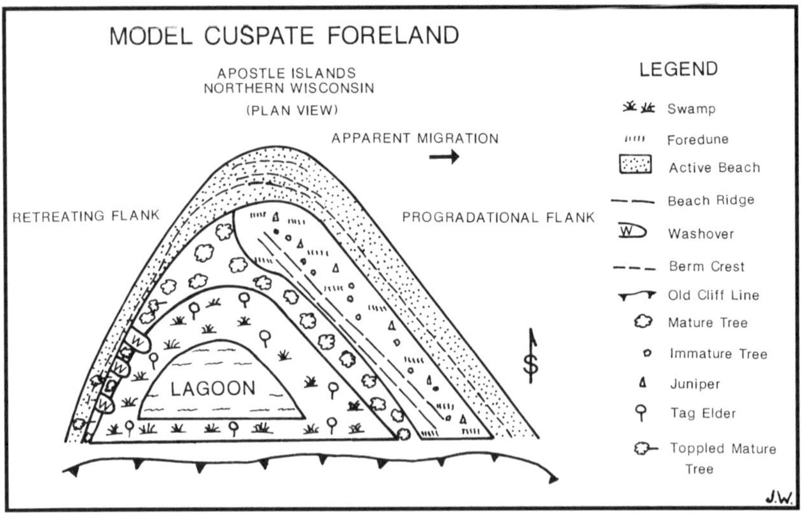

Figure 57. Diagram of a model cuspate foreland in the Apostle Islands (after Engstrom, 1985). South is toward top of diagram. In these islands, currents move sands generally north to south, and sands migrating along the east and west islands' coasts converge near the southern tip to create a cuspate foreland.

Twin Island. The highest point is a little above the elevation of 650 feet, just about 50 feet above lake level on a small hill at the south end of the island. This small hill may possibly be a drumlin; it is elongate in a northeast direction and parallel to the direction of ice movement. Glacial erratics (rocks carried great distances by glaciers—here, erratics are from Canada and the Lake Superior Basin) can be found along the entire shore of the island. These remained after finer materials were washed away by waves; thus, the sizes of the erratics are greatest in areas exposed to high-energy waves. At the north end, large boulders and cobbles of basalt, granite, and gneiss may be found. Near the sandspit, smaller gravels of various rock types may be seen.

The cuspate foreland (see Fig. 57) near the camp area and public dock is unique—it is one of two (the other is on Stockton) not found at the southerly end of an island. Here, the presence of the cuspate foreland is the result of unusual bathymetry of a shallow shelf that extends northward of the sandspit and continues to include all of South Twin and the northern half of Rocky Island. The low relief present on this immense shelf (only about 20 feet) indicates that the shelf was planed by wave and glacial action. Engstrom's (1985) data indicates that the spit has been decreasing in size over the past decade.

Common rocks found in the region of the Apostle Islands include igneous rocks—basalt, gabbro, granite, andesite, and diorite; the sedimentary rocks—arkose, quartzose sandstone, conglomerate, breccia, limestone and dolostone, concretions, and banded iron formations; and the metamorphic rocks—schist, slate, gneiss, quartzite, and metabasalt; and the minerals talc, chlorite, native copper, quartz (crystalline, milky, and agate varieties), feldspar, olivine, biotite, and muscovite. The visitor here can find a number of representatives from the above list along the gravel beaches of Rocky and South Twin islands.

Stockton Island

Map Reference—Stockton and Oak Islands 7 1/2' Quadrangles

Stockton Island (Fig. 58) is a favorite for sail boaters, day-users, campers, and hikers. Preliminary studies of the island were done by Stadnyk, Verch and Goetz (1974) and by Anderson and others (1980). The central location within the archipelago, the inviting sand beaches and shelter of Presque Isle Bay, the attractive hiking trails, and the beautiful beaches at Julian Bay make this island one of the most heavily visited. By coincidence, the island also contains major geological attractions. Balanced rocks, sea cliffs, glacial grooves, joints, a tombolo, a major quarry, and diverse sedimentary structures are available to both boater and hiker. Many of these lie in areas of the island that are not frequented; others lie in frequented areas but often go unnoticed or unrecognized for what they are. Several good days can be spent on Stockton Island with the geology alone.

Stockton is a large island, over 10,000 acres, and the Chequamegon Sandstone underlies all of it. Despite its size, relief is not great and its

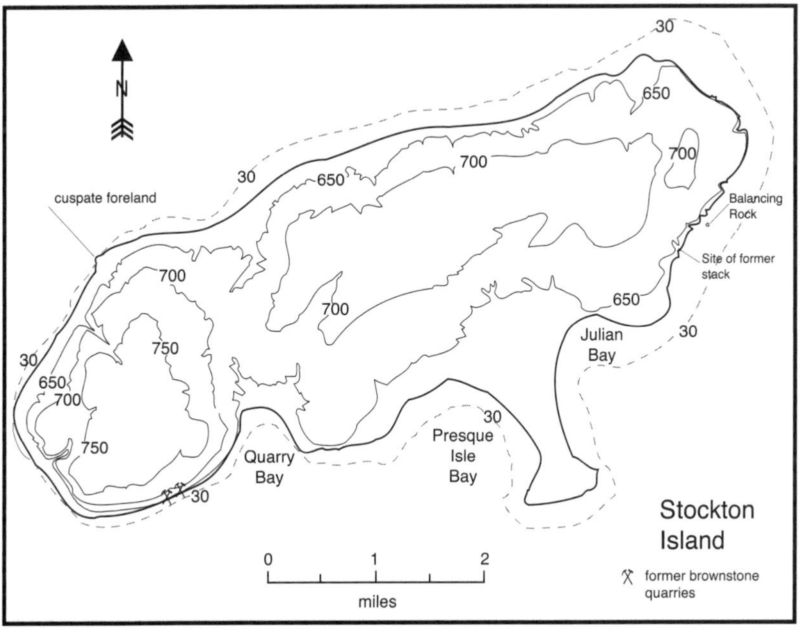

Figure 58. Map of Stockton Island showing general elevations (feet) and bathymetry (dashed lines).

highest point, about 800 feet, occurs near the southwest end. A prominent finger juts out from the southeast corner to form Presque Isle Bay. The finger, Presque Isle Point, occurs because wave-induced currents built up sand from the main body of the island southward to a second small island until the two were joined as "Stockton Island." The sand connection is termed a tombolo. Collie (1901) was probably the first investigator to refer to this particular tombolo. A later and detailed study was made by Coffin (1977). She referred to the junction as a "double tombolo" because two separate sandspits had migrated toward and eventually reached the second small island.

Coffin (1977) estimated that the tombolo began to form about 5,000 years ago and was largely completed 2,500 years later. In brief, the lake level was around five to ten feet higher about 5,000 years ago. The sandbars were built by currents that carried sand southward from the larger island. Eventually these underwater sandbars contacted the smaller island. As the lake level continued to drop, the tops of the sandbars were left above the water. Storms and possibly winter ice masses pushed some of the sand into ridges—the present beach ridges seen north of Julian Bay (Fig. 60). The lake waters continued to drop, thus leaving the beach ridges above the influence of storms or ice masses. Wind distributed some of the sands into dunes, and vegetation eventually established itself on the ridges and dunes, stabilizing them. Coffin cites accretion of sand dunes and stabilization by plant communities as important factors in allowing the tombolo to persist into the present day. As noted by Coffin (1977), the tombolo here is a delicate and transient feature—one that is slowly removed when the lake levels rise. Kraft (1982) made a later study of bank erosion at the Presque Isle campground.

Engstrom (1972) made a collective study of the modern depositional sandforms and linked the wave patterns generated by winds to the preferential occurrence of sandforms at the southern ends of the Apostle Islands. In the Apostle Islands, prevailing winds from the east (in May, June and August) and from the west (September through April and most of July—Coffin, 1977) dictate that sand will move in a southerly direction. Sandspits form at the southerly edges of those islands from which eroding glacial drift can supply a continuously abundant supply of sand (see Fig. 62). Islands without abundant glacial till (such as Eagle, Gull and North Twin) cannot support large sandforms.

The "singing sands" deserve special mention. "Singing sands" are those that yield conspicuous and sometimes musical sounds when the grains are abraded past one another. This usually occurs when the foot of a hiker is pushed through the sand to produce abrasion. Singing sands of the Great Lakes area were reported by Richardson (1919) from beaches along Lake Michigan. A thorough review of the characteristics of sound-producing sands was made more recently by Lindsay and others (1976). Lindsay and others divided such sands into those that emit high-pitched sounds (variously called "singing," "squeaking," "barking," "whispering" or "whistling" sands) and those which emit low-pitched sounds ("booming

Figure 59. Aerial photo of tombolo that attaches Presque Isle to the main Stockton Island taken between 1983 and 1987 (photo from National Park Service, Apostle Islands National Lakeshore). Presque Isle Bay, dock and entrance to campground is to left; Julian Bay is to right.

Figure 60. Aerial photo of Julian Bay showing wetland marsh and old parallel beach ridge strand lines to left taken between 1983 and 1987 (photo from National Park Service, Apostle Islands National Lakeshore). The Julian Bay beach with its white "singing sands" is one of the most beautiful in the islands. The deeply incised bay traps sand in an area such that it is reworked into smooth, rounded grains of quartz-rich sand.

sands"). Those at Julian Bay are characteristic of the former. Singing sands are those that have medium-sized grains that are well rounded and nearly spherical and which are comprised of grains that are all about the same size (well-sorted sands). Nuhfer and Hodgson (1987, 1988) used thin section microscopy (Fig. 61) and sieve analysis to examine sands that we collected in 1986 from numerous spits and beaches of these islands. Although the majority of sands in the islands are rich in potassium feldspar, medium grained, and well-sorted, the sands from Julian Bay and from the north end of Bear Island are significantly different in both shape and composition. They are more rounded and contain much less feldspar.

Thus the "singing sands" reported by hikers along Julian Bay have characteristics of singing sands and are truly distinct from those of other beaches in the islands. The characteristics of the Julian Bay sands probably arise from the fact that the sands have been trapped along this long concave stretch of beach and have been exposed to waves for long periods of time rather than swept into the deep areas of the lake after a few seasons. Consequently, the wave action has prolonged the grinding of sand grains upon one another, has worn away the less-resistant feldspar, and has resulted in a sand which is nearly 90 percent quartz, 8 percent feldspar, more well-sorted, more well-rounded, and probably smoother than sands found elsewhere on the islands (Fig. 61). As noted by Richardson (1919), the upper wave limit on the beach is about the upper beach limit for those sands with a propensity to "sing." As a general rule, the drier the sand, the more easily the characteristic "singing" is produced.

Figure 61. Singing sands from Julian Bay as seen under the binocular microscope (left) and in thin-section viewed through a polarizing petrographic microscope (right). The grains seen at left are smooth, well sorted, and well rounded. The black white and gray grains in right are almost all quartz; the plaid grain at top is a crystal of feldspar.

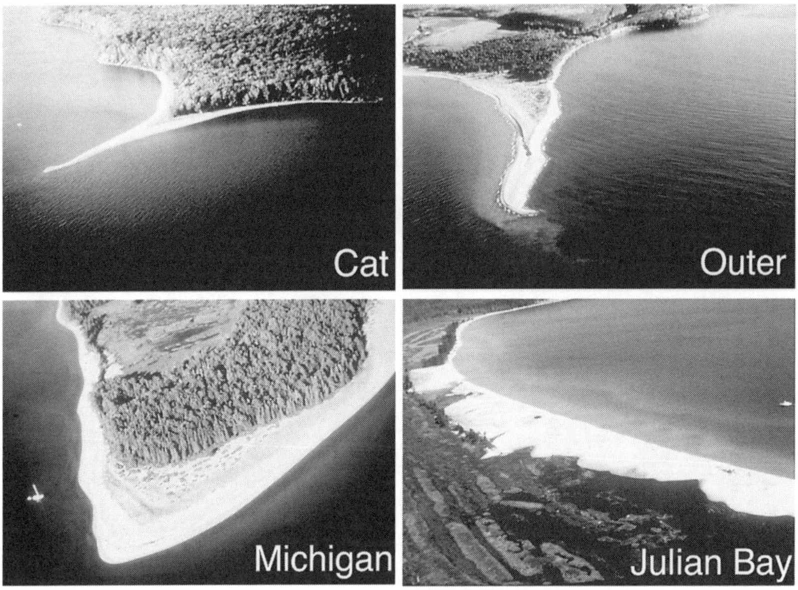

Figure 62. Aerial views of major sandforms in the Apostle Islands taken between 1983 and 1987. Wetlands created in cuspate forelands (see Fig. 57) and southerly-migrating sand can be clearly seen with the Outer and Michigan sandspits. The sandspits at Cat, Outer, and Michigan Islands are made of sand that makes a general one-way trip to depths below wave base. In contrast, the Julian Bay enclosure keeps sands exposed to waves for longer residence times. As they are moved back and forth, Julian Bay sands become more rounded and less resistant particles wear away (photos from National Park Service, Apostle Islands National Lakeshore).

Two trails lead along the tombolo: the Tombolo Trail and the Julian Bay Trail. The Julian Bay Trail is a good place to begin a tour of the Stockton Island tombolo. This trail leads through a forest dominated by evergreens and large yellow birch. It soon reaches an overlook that offers a scenic view of the wetlands (Fig. 63) within the tombolo.

You can walk northward along Julian Bay, see areas where storm waves have recently breached the sand ridges, and recognize that the high ridges are now being eroded. The multiple ridges present can be clearly seen from the ground and even more clearly from aerial photographs. If you continue hiking northward, you eventually reach bluffs of sandy glacial till. The gradual change of beach material from cobbles at the base of the till to pebbles, and sand in the bay itself certainly shows that sand is yet being supplied from the till to the lower beaches. While occasional storms destroy the ridges and remove the beaches, constructive processes rebuild the beach. As the lake waters continue to rise, the constructive processes slowly are being outstripped by the erosive ones. Beyond lowering the lake level, little can be done to stop this natural process. However, the life span of this exquisite area can be extended by remaining on the paths or on the beach itself and

Figure 63. View of wetlands within tombolo between Presque Isle Bay and Julian Bay. Cuspate forelands on other islands also harbor scenic wetlands.

not climbing over the foreshore ridges at Julian Bay or at the campsites. This will help to preserve the vegetation that stabilizes the dunes and ridges and slows their erosion. If you choose to walk northward, access will eventually be gained to the "Tombolo Trail," which winds into the woods and across bridges through the lush wetlands and back along the campsites.

An alternative, the Anderson Point Trail, begins with a hike to Julian Bay via the Julian Bay Trail and then turns right toward the rock ledges. This right turn will soon bring you to a spectacular view of an east-west trending joint (Fig. 64). The joint may be followed for some distance, and examination of the smooth straight nature of the break will aid you in identifying joint faces on exposed rock elsewhere. The rock lakeward of the joint is slowly separating and slipping toward the lake. As you proceed farther south, boulder and cobble beach rubble is found lying on top of wave-cut benches. Pounding waves and stresses of winter ice are slowly but relentlessly removing these rocks from the island. When waves are strong, you may stand on some of these boulders and feel the pulse or impact of the waves. Realize that each pulse is a microscopic movement of the rock that, in time, will move the rock lakewards. The smaller cobbles are generally flat, a form common to beach rock. Some show asterisk-like impact marks (Fig. 65), which bear testimony to the force with which they are hurled against each other by strong waves.

Still farther south, a rock surface shows excellent glacial striae (Fig. 66) that formed when rock that was carried in the base of the glacier gouged into this surface. The compass orientation of the grooves (in this case to the northeast) was governed by the direction from which the ice advanced. As you hike along the rocks, you will cross cliffs where old stream channels are exposed as discontinuous lenses or pods of sandstone. Cross-beds and joints are visible at many places between here and

Figure 64. Joint in Chequamegon Sandstone south of Julian Bay. Joints are linear breaks in the rock along which there has been little or no movement. The impact of waves and the wedging of ice are slowly moving the rocks on the left side of the joint toward the lake.

Figure 65. Impact marks in cobbles along Presque Isle Point. The asterisk-like breaks are produced when powerful storm waves hurl cobbles and boulders against one another.

Figure 66. Glacial striae on Chequamegon Sandstone at Presque Isle Point.

the visitor center. You can also see a few conglomerate beds, molds of clay galls, and even a few ripple marks, which exist as faint, almost questionable, impressions. The wave-cut cliff of the joined smaller island lies at the base of the overlook. As you follow the trail out onto the beach at Julian Bay, you can clearly see the old wave-cut cliff in the Chequamegon Sandstone. This was the northern shoreline of the smaller island prior to being joined by the sand tombolo.

The following features described at the unnamed point north of Julian Bay and east of Trout Point are accessible by hikers. However, in accord with the nature of this guide, we'll assume that the outcrops will be accessed by means of a small boat.

If the water is calm as you prepare to leave Presque Isle Bay, take your boat into the shallows around the campsites and observe the nature of the ripples forming in the sand. Like many ripple marks that you will see in the rocks on Stockton Island and Devils Island, these are ripple marks formed by shoaling waves. Many of the ripples branch or "bifurcate." Such branches are indicative of wave ripples. It is helpful to see some modern ones in order to compare these to ancient ones seen preserved in rock.

This guide will outline a route out of the bay, eastward around the rocky point (shallow shelves and boulders abound here, so stay a respectable distance from shore) into Julian Bay and northward along the east coast and around the entire island.

The unnamed promontory (accessible in good weather) at the north end of Julian Bay consists of unusually light-colored sandstones. The rocks show a number of geological features. Ripple marks are usually rare in the Chequamegon Sandstone, but here you can see several varieties along the wave-cut bench: undulatory ripples (Fig. 67) characteristic of slow currents, interference ripples (Fig. 68) characteristic of wind-generated waves in shallow ponded water, and current ripples (Fig. 69) all within a short distance of one another. The undulatory ripples here are particularly interesting because they were deposited along the edge of a sandbar where the currents rippled the channel bottom and the inclined sides of the channel. Modern analogues to these rippled surfaces can be seen at the mouth of the Sand River where it empties into Sand Bay. Figures 67-69 were taken in the 1986 and 1987 field seasons. We have since returned to that site to find even the large boulders and slabs moved by powerful waves. Visitors should expect to find our photographed slabs in other orientations and likely new features exposed.

Along the cliffs and broken edges of boulders are excellent examples of cross-bedding, particularly foreset beds and scour cross-beds. In many places, clay galls have been weathered from these sandstones, leaving their characteristic shallow cavities (Fig. 70). Several glacial grooves can be seen, particularly on the northern end of the bench.

After leaving the promontory, you proceed along a coastline of Chequamegon Sandstone that becomes increasingly rugged and interesting. Stacks come into view. Those closest to shore have been most recently isolated from the main island by erosion along closely spaced joints. Some of the nearby cliffs to the north exhibit Liesegang bands (Fig. 71).

Farther north, some of the upper ledges are brecciated (shattered into irregular fragments) from the stress imparted by overriding glacial

Figure 67. Undulatory ripples in Chequamegon Sandstone on Stockton Island. The undulatory ripples here are particularly interesting because they were deposited in a shallow channel that flowed across a sandbar where the currents rippled both the channel bottom and the sides. That the joint present at that boundary did not move or shift the rippled surfaces can be determined by carefully examining the continuity of the ripple forms across the joint.

Figure 68. Interference ripples in Chequamegon Sandstone on Stockton Island. These are produced in shallow, standing water where two sets of wave ripples result from two distinct wind directions. The wind blows from one direction and generates common wave ripples; when it later blows from a second direction, the resulting ripples are superimposed on the first set.

Figure 69. Casts of lingoid ripples in Chequamegon Sandstone on Stockton Island. Lingoid ripples occur in sand found beneath water that flows strongly in one direction, such as in a stream or tidal channel.

Figure 70. Flattened cavities, which result from erosion of clay galls or shaly rock fragments that were incorporated into sand carried by the Precambrian streams.

Figure 71. Liesegang bands are colored bands formed by stain from iron diffusing through the rock via ground water. The bands sometimes look like cross-beds but upon close inspection can be seen to cut across true bedding.

ice. A small waterfall spills out of a U-shaped stream valley nearby. Up the coast, you can see a small window where a pillar supports an open arch. Continuing even farther northward, you find the Balancing Rock (see back cover), a stack and a labeled feature on the Stockton Island Quadrangle map. The Balancing Rock consists of a series of channel sands; the upper layers facing the lake have been brecciated. An old photograph of this rock appeared in a Prentice Brownstone Company advertisement published in about 1892. This feature has changed perceptibly over a century but remains relatively stable in comparison to other stacks found here.

There are a number of rockfalls along the shore. Some of the Chequamegon beds become shaly, thinner, and jointed. This combination of features gives rise to a number of wave-sculptured features, including reentrants and a series of tiny sea caves. Thin beds of conglomerate in the cliffs farther north can be traced for some distance. Behind one small stack devoid of vegetation (informally called "Lucy Rock"—named for "Lucy" on the basis of an old name carved into the rock nearby), a deeply eroded joint may be entered for over 100 feet with a small boat. When the light is correct, brilliant shades of green shine from within the joint, where the sunlight plays over mosses that cover the joint faces. The stack photographed from inside this joint in our 1986 field work was gone by 1996 (Fig. 72), which is a testament to the ephemeral nature of stacks and other rock features found here, especially on the northerly edges of the islands. While completing this new edition in January 2004, this photo gave pause to thoughts about the relative strength of a small boat compared to that massive piece of rock. It was abundantly clear why the events of removal of such features have no eyewitnesses.

Toward the northeastern tip of the island, the shoreline changes to an increasingly jagged character, as ridges protrude out into the water. This "saw tooth" shoreline, caused by strong waves from the northeast eroding along joints in Chequamegon Sandstone, is marked by additional stacks, some clearly bounded by straight flat joint faces. Wave-cut platforms are present, and in one area, a sea cave rests upon a higher platform. In good weather, you can land a boat and walk through this short sea cave.

Your route continues around the north end of the island to Trout Point along a shoreline littered with many boulders and rockfalls. The northern and western shores of the island are mostly covered by vegetation and glacial till. Some very narrow beaches occur at the foot of the bluffs. On the northwest edge of the island, where the shoreline swings from southwesterly to more directly south, is a small cuspate foreland. After being transported along the long northern shore of the island, some sand accumulates here in the shelter afforded from the southwesterly longshore currents. Engstrom (1985) notes that the foreland's beach retreated about 1.8 feet between 1980 and 1983. Little rock other than dislodged boulders is seen until one reaches the southwest edge of the island, where some outcrops of Chequamegon Sandstone again appear. Farther ahead at Quarry Bay, a dock is available as well as campsites and a

Figure 72. The former stack ("Lucy Rock," left) viewed in 1986 looking eastward from a boat paddled into an opened joint. The straight edges mark planar joint faces. By 1996, (right) the stack had been swept away by storm waves and ice.

vault toilet. A trail leads up to the quarry, which was owned and operated by the Ashland Brownstone Company between 1889 and 1897. Over a million cubic feet of stone were removed between 1892 and 1896 (Buckley, 1898), and the opening is awesome even now, overgrown with vegetation.

From Quarry Bay, you can continue east, past a few low cliffs consisting of mottled cross-bedded sandstone, on into Presque Isle Bay.

York Island

Map Reference—York Island 7 1/2' Quadrangle

Collie (1901) was one of the earliest writers to recognize York Island as the product of a tombolo joining two smaller islands. Strzok (1981) notes that a survey done by a British lieutenant in 1824 showed two islands, the eastern one named York (the namesake of the present island) and a smaller western one called Rock Island. If Strzok's referenced date is valid and the lieutenant's map is a true representation, then it is an amazing fact that the tombolo (Fig. 73) formed in less than seventy-five years!

The York Island 7 1/2' Quadrangle map indicates a shelf about 20 feet below the present water surface. The shelf surrounds the island and connects with the mainland at Point Detour. This shelf is less than 6 feet below the water around the island and apparently allowed the tombolo to develop on a bedrock platform. Submerged boulders lie in the shallows and possibly move about the shelf as result of ice and storms. Care is needed when approaching this island. The northwestern end of the island, especially along the north shore, provides an extensive exposure of channel sandstones. A walk over the rocky platform reveals many con-

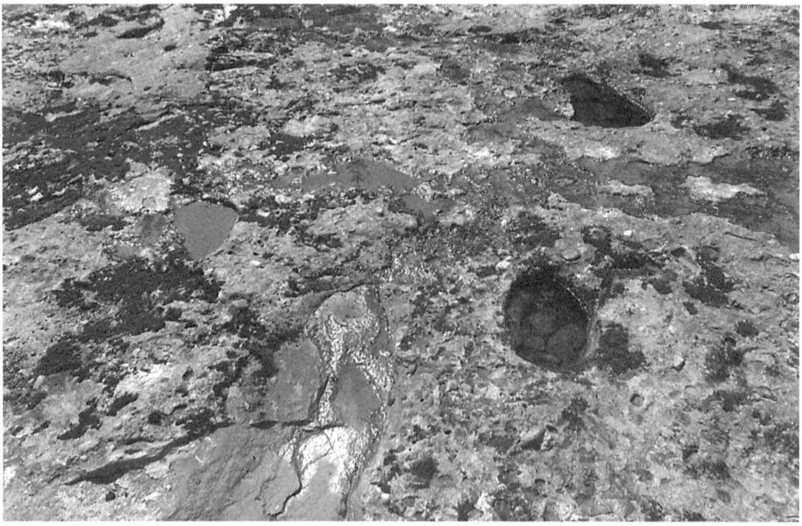

Figure 74. Potholes carved in Chequamegon Sandstone on wave-cut platform at York Island. High waves stir tools around the insides of cavities left by weathered-out cobbles. As they are stirred, they abrade the sides and bottom of cavities, slowly smoothing and enlarging them.

Figure 73. Aerial view of the tombolo on York Island (photo from National Park Service, Apostle Islands National Lakeshore).

glomerate exposures along bedding planes. The conglomerates are thin, possibly only a few pebbles thick, and were deposited when floods swept down the old Precambrian rivers, leaving an eroded surface covered by gravel that was then overlain by sand as the floodwaters waned. In some places, lake waves have used hard tools (igneous and metamorphic gravels carried in by glaciers) to erode small potholes (Fig. 74) into the sandstone bedrock. Between 1986 and 1996, we could see areas where bedrock at the water level had been forcefully broken, almost certainly the result of immense pressure exerted by thick winter ice.

The pleasant sandy beach along the central portion of the northern shore of York Island is a good place for anchorage and landing if you watch out for occasional boulders. Farther east along the north shore are more bluffs and cliffs of Chequamegon Sandstone. Accessible, scenic and fun to climb on, they exhibit faces that display intricate cross-bedding and convoluted laminae. The rocks extend a short distance down the western shore, where they become covered by glacial debris and brush. The southeastern tip is the site of a small sandspit; another that Engstrom (1985) notes has been receding in recent years.

Sheltered Water Tours Along the Mainland

Bark Bay to Little Sand Point

Map References—Bark Bay, Cornucopia, Squaw Bay, Sand Island 7 1/2' Quadrangles

The prevailing winds are generally from either the west or the northeast. The shape and orientation of the Bayfield Peninsula dictates that one side or the other of the peninsula will be in the lee of the winds and therefore more sheltered. These provide areas where suitable small boats and sea kayaks (not canoes or johnboats!) may gain access to first-rate scenery and geological features without having to cross distances of open water. To keep this guide at a reasonable length, we have not repeated here the geological details of the formations and processes of forming caves, beaches, etc. These are indexed and covered elsewhere in this guide.

Even in sheltered waters, care must be taken with the usual precautions of wearing lifejackets, notifying the National Park Service of your plans, and arranging to notify someone of the completion of your trip. The route here, with few exceptions (and these will be noted), is very close to shores that can be climbed if a bad weather emergency should arise. Much of this area is private property so one should land on shore only if a situation requires it. This trip starts from Bark Bay and, depending on weather, can end in Cornucopia or at the National Park Service dock at Little Sand Bay. If two cars are available, you may wish to leave one at the far end or arrange for transportation back to Bark Bay.

Bark Bay is a beautiful, narrow, sheltered bay west of Cornucopia. The access road traverses wetlands near the bayhead and then passes up the west side of the bay along stands of gigantic birch trees to a good public boat launch (marked on Fig. 1). The sunsets illuminate the cliffs of Orienta Sandstone on the east side of the bay, causing these reddish rocks to literally glow above the water. A tour along these cliffs in evening, even if you remain within the bay itself, is very worthwhile. From the boat launch, outcrops of the Orienta may be observed at Bark Point. The upper beds near the outer tip of Bark Point are brecciated from the stresses of glacial ice. The rocks show cross-beds, and bedding planes show lingoid ripples; both forms reveal the northeastward current flow of the rivers that deposited the Orienta Sandstone. Depending upon your boat, you may wish to hug the shore all the way down to the bayhead and then come up the east shore. By going this route, you will pass glacial tills and then come upon the sandy bayhead bar, which has accumulated the sands brought into the bay by the Bark River. This bar is similar in structure to the barrier bar at Big Bay State Park on Madeline Island. The sandy beach along this bayhead bar is one of the most beautiful beaches in the area. Behind the bar lie extensive wetlands. Near the eastern edge of the beach, a shallow slough of the Bark River enters the bay from the wetlands. If you are willing to walk your small boat through the few inches of water at the outlet, the slough deepens enough to allow use of oars or paddles, and you can travel for some distance into these attractive wetlands. Leaving the slough, you will eventually swing northwards up the east shore and pass more glacial drift until you reach the southernmost outcrops of the Orienta Sandstone emerging from beneath glacial drift.

The sandstones exhibit a variety of features, including channel sandstone beds, sandstone reentrants, recent rockfalls, older tumbled boulders, and intricate cross-beds which invariably dip to the northeast. Some gulls have taken residence among these cliffs. Above are the exquisite stands of birch from which the name "Bark Bay" surely must have been derived. By continuing up the east side toward Lake Superior, you encounter increasingly rugged coast. Joint openings become more pronounced and, near the tip, a feature occurs that merits the name "balancing rock" (Fig. 75) much more than does the formally named stack at Stockton Island. An arch (Fig. 76) is present that rivals the beauty of other arches on the islands. As you round the peninsula (Roman Point) into the main lake, Eagle Island comes into view. Ahead, the shoreline again swings southward into Siskiwit Bay. Spectacular outcrops of the Orienta continue south along the west side of Siskiwit Bay.

In just a few places, you may note round scars from drill holes traversing the thickness of the cliffs. These drill holes were not made in the days of the historical "brownstone quarries," but instead were drilled more recently with modern rotary drills. This rock was shot from the cliffs, loaded into barges and used for dock cribbing.

You can proceed along the shore, past the outcrops, and eventually along the shore of Siskiwit Bay toward Cornucopia. The west side of Siskiwit Bay contains many attractive rock terraces, which are easy to land on with a small boat. The beach at the head of Siskiwit Bay west of Cornucopia is extensive and attractive. Cornucopia itself has a good harbor, shops, and several old fishing boats displayed in drydock. There are also beautiful broad sandy beaches on the east side of Cornucopia. The sands in Siskiwit Bay were carried in via Siskiwit Creek and Lost Creeks.

If water conditions are not good, you should remain in Cornucopia because the tour farther toward Little Sand Bay requires good weather. As you proceed from Cornucopia, the Orienta Sandstone again appears on the east side of Siskiwit Bay, and outcrops continue around Squaw Point into Mawikwe Bay. (If you are using older maps, realize that the name "Mawikwe" has since been substituted for "Squaw" by the National Park Service on many features here—older maps and charts will show "Squaw Bay," etc.) Just within Mawikwe Bay at the eastern base of the point is the contact between the Orienta and the overlying Devils Island Formation. It is scarcely exposed, and the narrow strip of beach along Mawikwe Bay covers the shore for some distance ahead. Only under good weather conditions should you proceed ahead to the Mawikwe Bay cliffs and caves. The sea caves are most beautiful features, and in places rival those of even Devils Island. However, there are no landings, and the cliffs there cannot be climbed. A boat caught against these cliffs in rough water is in serious trouble.

Figure 75. Balancing rock of Orienta Sandstone at northeast end of Bark Bay. Rockfalls have isolated the rock, and beds have been eroded deeply at the base to produce the narrow neck of the pedestal that supports the mass of channel sandstones above. Photos: lower, taken in 1987 and upper, taken in 1997. Changes are mainly in removal of boulder rubble to the left of the stack, probably by waves.

Figure 76. Roman Point Arch formed in Orienta Sandstone at northeast end of Bark Bay.

The caves themselves are exquisitely carved (Fig. 77). Some are large enough to enter with a small boat, and the intricate network of pillars and arches is accentuated by their color. The best time to approach these is in early evening, when the western sunset illuminates the interior recesses and accentuates the varied colors. The Mawikwe Bay caves illustrate well the fact that geological processes are often rapid and that uniformitarianism (uniformity of process) is not equivalent to gradualism (uniformity of rate). The cliff feature "Elephant Rock" shown in figure 77 (lower) had probably persisted for decades. The name "Elephant Rock" is at least sixty years old, but as of late summer of 1987 (Fig. 77 upper), "Elephant Rock" existed no more. Tom Bredow of the National Park Service, through conversations with boaters and photographs taken by visitors, placed the time of collapse of Elephant Rock as late July or very early August, 1987, just after unusually severe cloudbursts caused large waterfalls and erosional events at several cliff areas.

In winter when Lake Superior freezes solidly, one can walk into the Mawikwe Bay caves and find a variety of ice formations (Fig 78). Ice is a powerful erosive force. In addition to the incredible forces generated at the shore level by Lake Superior's ice, both spray sent up by waves prior to lake freezing, spring waters seeping into cold air, and additional weight of any snow or freezing rain accumulate on cliff faces (Fig 79).

If you look carefully among the caves, you will note that in one area a joint has been weathered and widened deeply into the rock (Fig. 80). It is possible to enter this joint in a small boat and proceed into it for a long distance. Near sunset, varied green colors from mosses seem to glow and

Figure 77. Mawikwe Bay (formerly "Squaw Bay") Caves in the Devils Island Sandstone near Cornucopia are particularly beautiful in summer sunsets. Photos: lower, taken in summer of 1986 shows the "Elephant Rock" arch that was depicted on local postcards. Upper, taken in summer of 1987 shows the disappearance of Elephant rock that took place during a massive rockfall.

Figure 78. Mawikwe Bay Caves in winter of 1991. Ice columns, stalactites and stalagmites can be viewed here in winter. Note large broken column that reveals ice movement on lake below. Such ice movement is a strong force in shoreline erosion.

fluoresce; high above at the top of this canyon a natural bridge remains in place. The rocks reveal thin bedding and ripple marks, particularly ripple marks seen in cross section rather than along bedding planes. Lack of shelves and exposed bedding planes prevent you from seeing many ripples except for ripple casts in the roofs of caves and on broken boulders.

Ahead to the east is Sand Point, which has some outcrops of the Devils Island Formation and some slabs that show ripples. As you continue around Sand Point, you will enter Sand Bay, a broad bay characterized by long narrow beaches of sand. The first rock outcrops that you encounter at the west side of Sand Bay are sandstones of the Chequamegon Formation. Sand Bay gets its name from the extensive amounts of sand present, much of which is carried into the bay by the Sand River. A number of underwater bars are built around the mouth of Sand River, but with careful navigation, a small boat can enter. The sides and bottom of the sandy channel exhibit extensive ripple marks, similar to some seen in the billion-year-old sandstones at Stockton Island. The beach sands show beautiful lag deposits of alternating layers of light quartz-feldspar and dark magnetite-ilmenite. A cold, bubbling spring usually flows from this narrow beach. The east side of Sand Bay reveals some low outcrops of the Chequamegon Sandstone with cross-beds, channel cuts, and ripple marks. The Chequamegon is so similar in appearance to the Orienta (outcrops seen at Bark Bay are exemplary) that Thwaites (1912) once suspected that the two formations were the same. Ahead lies the park dock at Little Sand Bay.

Figure 79. Ice buildups along cliffs near Cornucopia add tremendous weight to the overhanging rocks. Coupled with the expansive force of freezing water, it is easy to see the kinds of stress the lake places on rock cliffs on the islands during winter.

Figure 80. Joint in Devils Island Sandstone among the Mawikwe Bay Caves. The joint has been enlarged by erosion and may be entered for some distance with a small boat or dinghy.

Lost in the Islands

"You can get lost in the Islands,"
he said
as he towed us to Cat.
"Some sailboaters get way out
and can't tell which island
is which.
Have you got lost yet?"
"Oh no," we said.
"We have our plastic placemat map,
a good compass
and a good eye."

"Oh yes," I said within.
"I have got lost here.
I cannot find my way back
to life as I knew it
before.
I cannot find the old me,
having left my fears back on Sand,
my impatience on North Twin.
I cannot find a whole me,
having left my heart on Devils,
my spirit with the eagle
on Otter.
Yes, I have got lost
in the Islands;
their spirit called to my spirit,
the purer part of me rejoined itself,
and I wait
here in Platteville at my desk
to find myself again
in the Islands."

Washburn to Houghton Point

Map References—Washburn and Long Island 7 1/2' Quadrangles

On the east side of the Bayfield Peninsula, a number of outcrops of the Chequamegon Sandstone are accessible. When west or northwest winds prevail, the water is usually calm adjacent to the east coast of the peninsula. From Bayfield, you can proceed south on Highway 13 and find outcrops of the Chequamegon Sandstone along the road cuts. At 3.2 and 3.6 miles south of the Bayfield city limits sign, two quarries (Pike's Quarry owned by Captain R. D. Pike and operated from 1888 to 1897) lie on the west (inland) side of the road. South of Washburn, a public park and boat launch offers good access to the eastern part of the peninsula. Our sheltered water tour begins here.

The stack-like "outcrops" in Van Deventer Bay near the boat launch are old concrete piers rather than rock, and these should be avoided because of obstructions around them. A true stack in the Chequamegon Sandstone, however, can be seen close to shore far to the southeast toward Ashland. To the east lies the peninsula of Chequamegon Point (see discussion under Long Island).

A small boat can stay close to shore and proceed north. In places, small basalt boulders of glacial erratics are present; the sandstone boulders have also been stained a dark color, so close inspection is needed to tell one from another. Past the docks at Washburn, the height of the bluffs begins to increase and better outcrops are seen. Stream channel environments for these sandstones may be recognized here by their variable thickening and thinning along shore. Along a few of the ledges are ripple marks. Steeply dipping foreset beds as well as most scour-and-fill crossbeds dip northeast, in accord with the flow direction of the streams that deposited these sandstones. The beautiful cliffs are on private property, so refrain from landing unless there is some emergency. The point of the mainland that protrudes furthest out into the lake, about three miles north of Washburn, is Houghton Point. About another mile to the north was the location for the main center of the old brownstone industry. The town of Houghton, which was located here, housed several quarries that were owned by the Washburn Stone, Hartley Brothers, and the Prentice Stone companies. Perhaps as much as 80 percent of all the brownstone came from these quarries near Houghton Point. The Prentice Quarry was by far the largest, and its stone was used for major buildings in ten states and forty cities (Buckley, 1898). Graded mill blocks in three grades (No. 1—free from blemishes and uniform in color; No. 2—with some white; No. 3—with more white), all square and sound, were produced along with ton (sound stones of irregular shape) and rock rubble. The Prentice Company constructed a large dock for some distance out into the lake and supplied it with steam loading equipment and rail tracks. While waves have destroyed most evidence of the dock above water, a line of stones marking the old cribs is still present. A sharp eye can detect a remnant of the railroad spur on the bluff above the dock remnants, by the compact-

Figure 81. Channel sandstones in the Chequamegon Formation at northern end of Houghton Point. Note the concave downward cut made by a stream channel and the subsequent filling of this channel with sand. This fill is now a sandstone bed that thins at the edges. This is a classic example of what geologists call a "cut-and-fill" structure.

ed piles of broken brownstone. Scant evidence of the huge sizes of the quarries is seen from the lake. However, the old quarries, today on private property, are immense pits now flooded with water. A spectacular channel cut-and-fill (Figure 81) may be found a bit farther to the north.

To the east, you can see Long Island and Madeline Island. The next rock outcrops lie ahead at Van Tassels Point, below the Pike's Quarry outcrops viewed on Highway 13.

References Cited

Abel, C. D., 1985, Petrology and sedimentology of the Jacobsville Sandstone (northern Michigan) and Bayfield Group (northern Wisconsin): M.S. thesis, University of WI at Madison, 229 p.

Allen, J. R. L., 1982, *Sedimentary Structures: Their Character and Physical Basis*: Elsevier, v. 1, pp. 441-443.

Anderson, R. K., Milfred, C. J., Kraft, G. J., and Fraundorf, W. J., 1980, Inventory of select Stockton Island resources for recreational planning: Final Report to National Park Service, University of WI at Stevens Point, 169 p.

Berkson, J. M., 1972, Microrelief of western Lake Superior: Ph.D. thesis, University of WI at Madison, 63 p.

Bona, L. J., 1990a, Geomorphology and Holocene geologic history of Long Island, Apostle Islands National Lakeshore, Wisconsin: in Geological Society of America Abstracts with Programs, March 1990, v. 22, Issue 5, p. 4.

Bona, L. J., 1990b, Geomorphology and Recent geologic history of Long Island, Apostle Islands National Lakeshore, Wisconsin: M.S. thesis, Madison, WI, University of Wisconsin at Madison, 219 p.

Buckley, E. R., 1898, *On the Building and Ornamental Stones of Wisconsin*: WI Geological and Natural History Survey, Bulletin 4, Economic Series n. 2, 544 p.

Burnham, G. M., 1930, *The Lake Superior Country in History and in Story*: Chapple Publishing Co, Boston, 464p.

Cant, D. J., 1982, Fluvial facies models: in *Sandstone Depositional Environments*, P. Scholle and D. Spearing, eds., American Association of Petroleum Geologists, Memoir 31, Tulsa, OK, pp. 115-138.

Cary, S. J., McDowell, P. F., and Graumlich, L. J., 1979, Soils and surficial geology of four Apostle Islands: WI Academy of Sciences, Arts and Letters, v. 67, pp. 14-30.

Clayton, Lee, 1985, Pleistocene geology of the Superior region, Wisconsin: WI Geological and Natural History Survey Information Circular, n. 46, 40 p.

Coffin B. A., 1977, Plant distribution in relation to geomorphic processes on Presque Isle Tombolo Stockton Island, Lake Superior: M.S. thesis, University of MN, 63 p.

Collie, G. L., 1901, Wisconsin shore of Lake Superior: Geological Society of America Bulletin, v. 12, pp. 197-216.

Connell, D. E., 1984, Distribution, characteristics, and genesis of joints in fine-grained till and lacustrine sediment, eastern and northwestern Wisconsin: M.S. thesis, University of WI at Madison, 227 p.

Davidson, D. M. Jr., 1982, Geological evidence relating to interpretation of the Lake Superior Basin structure: in *Geology and Tectonics of the Lake Superior Region*, R. J. Wold and W. J. Hinze, eds., Geological Society of America Memoir 156, pp. 5-14.

Edil, T. B., and Shultz, M. N., 1983, Landslide hazard determination along a shoreline segment: Engineering Geology, v. 19, pp. 159-177.

Engstrom, W. N., 1972, Spatial patterns in beach morphology and sedimentation in the Apostle Islands of northern Wisconsin: Ph.D. thesis, University of WI at Madison, 236 p.

Engstrom, W. N., 1973, A method of predicting beach orientation: Professional Geographer, v. 25, pp. 12-15.

Engstrom, W. N., 1974a, Areal variations in beach dimensions in Apostle Islands of Lake Superior: Yearbook of Association of Pacific Coast Geographers, v. 39, pp. 33-50.

Engstrom, W. N., 1974b, Beach foreshore sedimentology and morphology in the Apostle Islands of northern Wisconsin: Journal of Sedimentary Petrology, v. 44, pp. 190-206.

Engstrom, W. N., 1985, Recent beach erosion along spits and cuspate forelands of Apostle Islands National Lakeshore: Shore and Beach, v. 53, pp. 22-28.

Farrand, W. R., 1960, Former shorelines in western and northern Lake Superior basin: Ph.D. thesis, University of MI, Ann Arbor, MI, 226 p.

Farrand, W. R., 1969, The Quaternary history of Lake Superior: Proc. 12th Annual Conference on Great Lakes Research, pp. 181-197.

Gable, D. J., and Hatton, T., 1983, Maps of vertical crustal movements in the conterminous United States over the last 10 million years: U.S. Geological Survey Misc. Invest. Series Map MI 1315.

Halls, H. C., and Pesonen, L. J., 1982, Paleomagnetism of Keweenawan rocks: in *Geology and Tectonics of the Lake Superior Region*, R. J. Wold and W. J. Hinze, eds., Geological Society of America Memoir 156, pp. 173-201.

Halls, H. C., and West, G. F., 1971, A seismic refraction survey in Lake Superior: Canadian Journal of Earth Science, v. 8, pp. 610-630.

Hartford, F., and Tanner, W., 1976, Current Great Lakes shore erosion damage: Shore and Beach, v. 44, pp. 16-19.

Hinze, W. J., Wold, R. J., and O'Hara, N. W., 1982, Gravity and magnetic anomaly studies of Lake Superior: in *Geology and Tectonics of the Lake Superior Region*, R. J. Wold and W. J. Hinze, eds., Geological Society of America Memoir 156, pp. 203-221.

Johnson, M. D., 1980, Origin of the Lake Superior red clay and glacial history of Wisconsin's Lake Superior shoreline west of the Bayfield Peninsula: M.S. thesis, University of WI at Madison, 96 p.

Judziewicz, E. J., and Koch, R. G., 1993. *Flora and Vegetation of the Apostle Islands National Lakeshore and Madeline Island, Ashland and Bayfield Counties, Wisconsin*: The Michigan Botanist, v. 32, n. 2, pp. 43-189.

Kalliokoski, J., 1982, Jacobsville Sandstone: in *Geology and Tectonics of the Lake Superior Region*, R. J. Wold and W. J. Hinze, eds., Geological Society of America Memoir 156, pp. 147-155.

Keroher, G. C., and others 1966, Lexicon of geologic names of the United States for 1936-1960: U.S. Geological Survey Bulletin 1200, 4391 p.

Kowalski, W. L., 1976, Geology, soils, and recreational limitations of Bear Island, Wisconsin, Apostle Islands National Lakeshore: M.S. thesis, MI Tech. University, Houghton, MI, 199 p.

Kraft, G. J., 1982, Soil and recreational use on the Apostle Islands National Lakeshore: I. Soil suitability for trails and campsites; II. Bank erosion in the Presque Isle campground: M.S. thesis, University of Wisconsin at Stevens Point, 96 p.

Larsen, W. C., 1975, Vegetation ecology of Bear Island, Apostle Islands National Lakeshore: M.S. thesis, MI Tech. University, Houghton, MI, 91 p.

Leatherman, S. P., 1982, *Barrier Island Handbook*: University of Md., College Park, MD, 109 p.

Leith, C. K., Lund, R. J. and Leith, Andrew, 1935, Pre-Cambrian rocks of the Lake Superior region: U.S. Geological Survey Professional Paper, 184, 34 p.

Leverett, F., 1929, Moraines and shorelines of the Lake Superior region from Late Keweenawan to Late Cambrian time: Geological Society of America Bulletin, v. 72, pp. 1-18.

Lindsay, J. F. and others, 1976, Sound-producing dune and beach sands: Geological Society of America Bulletin v. 87, pp. 463-473.

Luetgert, J. H., 1982, The earth's crust beneath Lake Superior—An interpretation of cross-structure seismic refraction profiles: Ph.D. thesis, University of WI at Madison, 126 p.

Matteson, R. E., and others, 1973, The impact of the Apostle Islands National Lakeshore on the local government institutions of Bayfield County: Sigurd Olson Inst. Envtl. Studies, Northland College, Ashland, WI, p. 88.

McCabe, C., and Van der Voo, R., 1983, Paleomagnetic results from Upper Keweenawan Chequamegon Sandstone: implications for red bed diagenesis and late Precambrian apparent polar wander of North America: Canadian Journal of Earth Science, v. 20, pp. 105-112.

Meller, H. V., 1987, The erosional control of joint orientations in the Bayfield Sandstone Group of the Apostle Islands of Lake Superior: Sand, Devils, North Twin, and Outer islands: unpublished manuscript, University of Wisconsin at Platteville Geology Program Independent Study paper, 25 p.

Middleton, B. A., 1983, Sandspit vegetation and geomorphic processes in the Apostle Islands, Wisconsin: M.S. thesis, University of MN, Minneapolis, 89 p.

Morey, G. B., and Green, J. C., 1982, Status of the Keweenawan as a stratigraphic unit in the Lake Superior region: in *Geology and Tectonics of the Lake Superior Region*, R. J. Wold and W. J. Hinze, eds., Geological Society of America Memoir 156, pp. 15-26.

Morey, G. B., and Ojakangas, R. W., 1982, Keweenawan sedimentary rocks of eastern Minnesota and northwestern Wisconsin: in *Geology and Tectonics of the Lake Superior Region*, R. J. Wold and W. J. Hinze, eds., Geological Society of America Memoir 156, pp. 135-146.

Morey, G. B. and Van Schmus, W. R., 1988, Correlation of Precambrian rocks of the Lake Superior region, United States: in Harrison, J. E., and Peterman, Z. E., eds., Correlation of Precambrian rocks of the United States and Mexico: U.S. Geological Survey Professional Paper, 1241-F, p. F1-F31.

Mudrey, M. G., Jr. and Ostrom, M. E., 1986, Upper Keweenawan stratigraphy, chapter IV: in Mudrey, M. G., Jr., ed., Precambrian petroleum potential, Wisconsin and Michigan, Wisconsin Geological and Natural History Survey, Geoscience Wisconsin, v. 11, p. 23-25.

Myers, W. D., 1971, The sedimentology and tectonic significance of the Bayfield Group (Upper Keweenawan?) Wisconsin and Minnesota: Ph.D. thesis, University of WI at Madison, 269 p.

Need, E. A., 1980, Till stratigraphy and glacial history of Wisconsin's Lake Superior shoreline: Wisconsin Point to Bark River: M.S. thesis, University of WI at Madison, 45 p.

Need, E. A., and others, 1980, Shoreline erosion and bluff stability along Lake Michigan and Lake Superior shorelines of Wisconsin; Appendix 9, Douglas and western Bayfield counties, Wisconsin Point to Bark Bay: WI Geological and Natural History Survey, Shore Erosion Study Technical Report, 272 p.

North American Commission on Stratigraphic Nomenclature, 1983, North American Stratigraphic Code: American Association of Petroleum Geologists, Tulsa, OK, 24 p.

Nuhfer, E. B., and Dalles, M. P., 1994, Lake Superior's Apostle Islands National Lakeshore as a resource for geologic education: in Geological Society of America Abstracts with Programs, April 1994, v. 26, Issue 5, p. 56.

Nuhfer, E. B., and Dalles, M. P., 1987, The Precambrian Bayfield Group of the Apostle Islands—a showcase for braided stream features: (abs.) WI Academy of Sciences, Arts and Letters, 117th Annual Meetings Conference Proceedings, Ripon, WI, p. 19-20.

Nuhfer, E. B., and Hodgson, S. A., 1987, Petrography of the singing sands of Stockton Island and comparisons with other sands of the Apostle Islands National Lakeshore, Wisconsin, USA: Ninth Research Conference Proceedings of Apostle Islands National Lakeshore, Bayfield, WI, p. 8.

Nuhfer, E. B., and Hodgson, S. A., 1988, Petrography of the singing sands of Stockton Island, Apostle Islands National Lakeshore, northwest Wisconsin, USA: WI Academy of Sciences, Arts and Letters, 118th Annual Meetings Conference Proceedings, Menomonie, WI, p. 20-21.

Ojakangas, R. W., and Morey, G. B., 1982a, Keweenawan sedimentary rocks of the Lake Superior region—a summary: in *Geology and Tectonics of the Lake Superior Region*, R. J. Wold and W. J. Hinze, eds., Geological Society of America Memoir 156, pp. 157-163.

Ojakangas, R. W., and Morey, G. B., 1982b, Proterozoic sedimentary rocks (abs.): in *Geology and Tectonics of the Lake Superior Region*, R. J. Wold and W. J. Hinze, eds., Geological Society of America Memoir 156, pp. 83-84.

Reed, J. C., 1993, Generalized correlation chart for the Precambrian rocks of the conterminous United States: in Reed, J. C., Jr., and others, eds., Precambrian; conterminous U.S., Geological Society of America, The Geology of North America, The Decade of North American Geology (DNAG), v. C-2, pl. 7.

Reineck, H. E., and Singh, I. B., 1980, *Depositional Sedimentary Environments with Reference to Terrigenous Clastics* (2nd ed.): Springer Verlag, NY, 549 p.

Richardson, W. D., 1919, The singing sands of Lake Michigan: Science, v. 50, pp. 493-495.

Rose, W. J., 1988 Water resources of the Apostle Islands National Lakeshore, northern Wisconsin: U. S. Geological Survey, Water-Resources Investigations, WRI 87-4220, 44 p.

Scholz, C. A., 1984, Late and post-glacial sediment distribution in western Lake Superior from seismic reflection profiles: (abs.), Proc. 30th Annual Institute on Lake Superior Geology, pp. 61-62.

Schultz, R. A., 1976, Animal ecology of Bear Island, Wisconsin, Apostle Islands National Lakeshore: M.S. thesis, MI Tech. University, Houghton, 49 p.

Shetron, S. G., 1974, Soil investigation on Bear Island, Wisconsin, Apostle Islands National Lakeshore: Department of Biological Sciences, MI Tech. University, Houghton, MI, 7 p.

Stadnyk, L., Verch, R. L., and Goetz, B. A., 1974, Stockton Island survey: Northland College, Ashland, WI, 71 p.

Sterrett, R. J., 1980, Factors and mechanics of bluff erosion on Wisconsin's Great Lakes shorelines: Ph.D. thesis, University of WI at Madison, 389 p.

Strzok, D., 1981 & 1988 & 2003, *A Visitor's Guide to the Apostle Islands National Lakeshore*: (three editions 1981, 1988, with 2003 as the latest edition), Ashland, WI, Heart Graphics, Inc., 178 p. in 2003 edition.

Thwaites, F. T., 1912, Sandstones of the Wisconsin coast of Lake Superior: WI Geological and Natural History Survey, Bulletin 25, Scientific Series n. 8, 117 p.

Tyler, S. A., Marsden, R. W., Grout, F. F., and Thiel, G. A., 1940, Studies of the Lake Superior Precambrian by accessory mineral methods: Geological Society of America Bulletin, v. 51, pp. 1429-1538.

Van Schmus, W. R., Green, J. C., and Halls, H. C., 1982, Geochronology of Keweenawan rocks of the Lake Superior region: a summary: in *Geology and Tectonics of the Lake Superior Region*, R. J. Wold and W. J. Hinze, eds., Geological Society of America Memoir 156, pp. 165-171.

Watts, D. R., 1979, Upper Keweenawan and Lower Paleozoic paleomagnetism of the North American Craton: Ph.D. thesis, University of MI, Ann Arbor, MI 257 p.

White, W. S., 1966, Tectonics of the Keweenawan basin, western Lake Superior region: U.S. Geological Survey Professional Paper 524E, 24 p.

Wilson, L. R., 1931, Evidence of a lower level of Lake Superior: Science, v. 75, p. 75.

Wold, R. J., Hutchinson, D. R., and Johnson, T. C., 1982, Topography and surficial structure of Lake Superior bedrock based on seismic reflection profiles: in *Geology and Tectonics of the Lake Superior Region*, R. J. Wold and W. J. Hinze, eds., Geological Society of America Memoir 156, pp. 257-272.

Further Sources of Information

Wisconsin Geological & Natural History Survey 3817 Mineral Point Road Madison, WI 53705 Phone 608-262-1705, website http://www.uwex.edu/wgnhs/. For topographic quadrangle maps referenced in this guidebook and many additional geological publications.

United States Department of Interior, Apostle Islands National Lakeshore Route 1, Box 4 Bayfield, WI 54814 Phone 715-779-3397, website http://www.nps.gov/apis/, email APIS_Webmaster@nps.gov. For maps, charts, publications, information on docking, access, campsites, programs & activities, current research within the national lakeshore, weather and safety information.

Acknowledgements

The first edition of this guidebook would not have been possible without a number of contributors. The National Park Service (NPS) provided logistic and particularly outstanding emotional support for both the first and the most recent project. The first edition was made possible through a grant from Eastern National Park and Monument Association that was managed by Project Coordinator Merryll Bailey, Park Researcher at the Apostle Islands National Lakeshore. Merryll Bailey (NPS) and Dr. Al Swain suggested the need for a geological field guide and this led to a funded project proposal. Special thanks go to NPS personnel for help and discussions. We are particularly indebted to Art and the late Phoebe MacLeod of Washburn for their information on the history of the brownstone industry, for copies of documents, and for tours of the quarries near Washburn. Apostle Islands' rangers allowed us to proceed with our work in bad weather by towing our small boat to sites when needed. Ross Marine of Washburn performed a minor miracle on our thirty-year-old, and later, forty-year-old outboard motor and those special efforts enabled us to complete successful field seasons. University of Wisconsin—Platteville undergraduates were instrumental in several phases of this project. Harry Meller provided field assistance in 1986 and data on a study of joints that was performed during his internship in 1985. Scott Hodgson performed an excellent laboratory study of the beach sands that we collected from a number of islands, and cartography student Jeff Weymouth drafted maps and line drawings for the first edition. Colleagues Dr. Dwayne D. Stone, Dr. Phillip J. Gerla, Ken Shubak, and the late William A. Broughton are thanked for their reviews and helpful discussions, Dr. Stone's careful editing was especially appreciated.

The second edition was needed because the first edition of 2000 copies sold out years ago. A revised book was made possible by release time in the 1996 field season for the senior author from University of Colorado at Denver by Vice Chancellor Dr. David Groth and Chancellor Dr. Georgia Lesh-Laurie, who both valued administrators keeping active as academic researchers. Many things seemed to conspire against getting a manuscript completed afterwards—everything from multiple family tragedies to career relocations. As a result, a project to document a decade of change through photos of the Islands rocky lakeshores was simply wrecked by these disruptions. Life returned to normal enough to allow generating a product only in late 2003, and this book's illustrations benefit from being able to show the changes we documented. I'd like to thank personnel Neil Howk and Julie Van Stappen at Apostle Islands National Lakeshore and Rachel Shumsky of Eastern National for their extreme patience during the years that intervened before any product could be reality. Neil Howk was especially helpful for suggestions and edits in the revised edition.

Index